BECOMING A Dog Chef

BECOMING A Dog Chef

Stories and Recipes to Spoil Your Pup from America's Top Dog Chef

Kevyn Matthews

Turner Publishing Company
Nashville, Tennessee
www.turnerpublishing.com

Cover Design: Elina Diaz

Cover Photo/illustration: stock.adobe.com/Javier brosch

Interior Photos: Meredith Brown

Layout & Design: Elina Diaz

Becoming a Dog Chef: Stories and Recipes to Spoil Your Pup from America' s Top Dog Chef

Library of Congress Cataloging-in-Publication number: 2021936038
ISBN: (print) 978-1-64250-441-5, (ebook) 978-1-64250-442-2
BISAC category code PET004000, PETS / Dogs / General

Printed in the United States of America.

Table of Contents

Chapter 1

The Story of Greta

July 15, 2003, around seven thirty in the evening, on the A train to Far Rockaway. I was going to pick up my new puppy from the airport who was coming from Ohio on a cargo flight. "*Stand clear of the closing doors!*" I heard, as the doors to the subway shut and the train slowly started to move. I dozed off as the train rocked me back and forth on the way to our destination. I dreamed about my childhood dogs and how my dad loved them.

We had two Dachshunds, Heidi and Greta. Heidi was the reddish colored one who always seemed to get in trouble, and Greta was the black and tan one who seemed to do everything right. My father loved those dogs. They were like his little buddies: everywhere he went in the house, they followed, especially when he went to the kitchen! They never seemed to eat "dog food" or like it, so dad usually gave them what we ate for dinner on top of their food to make them eat it. Without the table scraps, you'd have to get down on your hands and knees and act like you were eating their food to get them interested in it.

They didn't exercise much, other than playing in the backyard, so they gained a significant amount of weight. They were literally eating four to five times what they should have been eating per day in table scraps, and definitely more during the holidays. After a while they became obese, and we lost Greta to a heart attack. After losing Greta, we tried to incorporate healthy options for Heidi, and even started walking her. *What can I do to make her live forever?* I wondered. While Heidi was making strides in a healthier direction, things started to take a turn for the worse. She refused to walk up steps, but I always carried her and never thought about it. After a while, she stopped walking. When we took her to the veterinarian, he said: "I have a treatment for this, but the medication will eventually stop working. Then there's nothing I can do."

My dad and I left the veterinarian's office and sat in the car in silence. I could hear the wind whistling as leaves breezed by our car on that winter afternoon. The sun started to go down, and as I stared at the orange clouds bouncing off of the calming blue sky, I could smell the crisp air through the crack in the window. I knew he was upset, as was I, but I didn't want him to see me cry. My dream of Heidi living forever was doomed by the reality of life's cycle itself. I was heartbroken, and I could feel Dad covering his sadness as I, too, was covering mine. He started the car and drove home, and neither one of us said a word. When we got home, we ate dinner, but Heidi refused to eat. She never did that. We offered her everything, but she just gave us this sad look and turned away, almost as if she was ashamed or hurt. When I asked her if she wanted to go outside, she seemingly perked up but couldn't walk. I picked her up and carried her outside and said, "Go on girl, pee." She didn't move. "Heidi?" I said, as though she didn't hear me. She turned her head, and when she looked at me, it almost appeared as if she had tears in her eyes. I felt her thoughts to me and heard, "I'm going to miss you." The look on her face, I'll never forget. I felt her sadness, as if she was going on a long trip and would never see me again.

"*Stand clear of the closing doors, please. Next stop, JFK Airport*," I heard as I yawned and woke up.

"Damn man, you been asleep this whole time?" said the man sitting next to me. "Yeah, I must have dozed off." I said as I sat up in the seat. "These trains do that sometimes. You know they say it makes us fall asleep because it soothes us like the mother's womb," he said as he smiled. "Where you goin'?" "I'm going to meet my new puppy! She's gonna be my new Doberman Pinscher," I said with excitement as the train came to its last stop.

"What are you gonna name her?" the man said. "Greta," I said with a smile. I grabbed her blanket, crate, and toys, and stood up to walk out when the man said, "Good luck with Greta!" I thanked him and ran toward the airport.

I had never been to JFK airport, so I had to come to an abrupt stop to figure out where to go. In the distance, I heard a woman's voice: "Sir?" I turned around so quickly that I dropped the toys on the ground. As I reached down to pick them up, an older woman walked toward me. "I'm sorry I startled you, but you look confused. Can I help you?" "Yes, hopefully you can! I don't know where I'm going, but I'm looking for the cargo area. I'm here to pick her up." This seemed to confuse her. "Who is she? People don't get picked up in cargo, sir." I explained, "I'm here to get my puppy Greta and they said she'd be in the cargo area!" I said. She got a big grin on her face and said "Aww!! We gotta go get Greta!" So I followed her as she ran down the hallway. "FOLLOW ME!" she screamed, as she dashed through the airport. It seemed like she was just as excited to get me to Greta as I was to see her. After running through the airport for about five minutes, she slowed down when we got to a set of gray double doors. They had a hazard sign on them and looked like they went to the runway. "Go through those doors and go to the right, the cargo area is in that building," she said. "Okay, thank you so much!" As she walked away, she turned around and said, "That dog is gonna be something special, I can tell." "Me too," I replied.

As I walked through those doors, I looked to the left and saw the planes sitting there. I felt like this moment was bigger than any airplane for me—this was a dream come true.

As I walked through those doors to pick her up, I controlled my excitement and asked the gentleman at the front desk if he could help me. "Excuse me sir, I'm here to get a package." He looked at his screen and said, "Nothin' here just yet, have a seat." As I sat down, I noticed a stack of magazines, so I picked one up and flipped through it. As I flipped through the pages, I thought about the first time I saw a Doberman and knew I wanted one. I was in the car with my dad on a cold Sunday morning getting yelled at for my bad grades. In the middle of his rage, he froze mid-sentence. "Look at that beautiful dog," he said as he stopped the car. "Look at her coat, man. Look at how strong and proud she looks when she walks." People were honking at us because we stopped moving, but he didn't care. "That's a real dog, man. That's a Doberman," he said with pride. "Look at her owner, he cares more about her than himself!" I stared at the owner and dog walking and could see the pride in both of them. "You don't have to teach them anything, those dogs will protect you with their lives." He then put the car in drive and slowly drove off. As I stared at them, I realized I wanted a dog just like that. (My father also forgot about the whole conversation about my bad grades.)

"Is there a Kevyn Matthews here?!" a man screamed. "That's me, sir," I said as I stood up. "Plane just came in and your package is here," he said, pressing the footswitch to operate the conveyer belt. Box after box passed, nothing. Then I see a box coming down the conveyer belt that says, "Live Animals" and it's *moving*! "What's *that*?" he said. "It's Greta!!" I exclaimed. I pulled the moving box from the belt and signed the paperwork. He said, "She's gonna be something special man." I just smiled and said, "I know." I had to switch her from the temporary crate to the one I brought with me, so I put everything on the ground. "Hi beautiful," I said. As I put the box down, it shook with excitement. I opened it and she walked past me to urinate, then jumped all over me with excitement. "Hey girl!" I said as I picked her up and hugged her. "You're gonna love this life!" I said as I put her in her new crate with toys. I put her crate on a luggage cart and pushed her through the airport. The train was empty when I got on it, but quickly got packed. I tried to stay quiet because animals aren't allowed on the subway. But as the train rocked back and forth to Manhattan, Greta started to cry. "Shhhh!" I said, but it didn't seem to help. I tried one more time, and that didn't work either. At this point, I start freaking out because people were watching. Against my better judgement, I opened the crate and put her on my lap. The response was the exact opposite of what I thought—everyone on the train wanted to touch her and wish me well with her. Getting out at my stop was like leaving your parents' house to go to college: sad goodbyes and best wishes for Greta from the late night (A) train to Far Rockaway.

When I stepped out of the train station at 96th and Lex at around eleven thirty in the evening, it seemed like Greta and I were the only ones on the block. It was a beautifully clear and starry night. I walked home as fast as I could, only to stop at the doorstep and sing Happy Birthday to myself. I couldn't wait to let her see her new home.

It took her a while to get used to the city—she came from a quiet farm and needed to acclimate to the street noise. I had to take her out late at night because the heavy foot traffic in the daytime startled her. I lived three blocks from Central Park, so I'd take her for walks after dark until she was okay with being around people. When she got older, I taught her to run off-leash in the park, and she loved the freedom of an open run. Once Greta became comfortable, there wasn't anywhere she wouldn't go. I took her into stores with me, all over Manhattan and the Bronx...we went everywhere together.

One afternoon, I was walking down Lexington Avenue with Greta and we passed by a butcher shop. The owner came outside and said, "Come here, I have something for her!" The store was layered with fresh meats and fish in refrigerated merchandisers. The owner, standing behind the counter with his blood-stained apron, reached over the counter with a ball of raw meat and said "Here ya go girl!" Before I could say a word, she gobbled up the whole thing. When we walked out of that butcher shop, she seemed to have a different pep in her step. We walked over five miles that day, and it didn't seem to affect her. When we got home and I went to feed her, she refused to eat the dog food. I thought maybe she wasn't hungry because of the walk. But the next morning, the same thing happened and I got a bit worried, so I took her to the vet. The vet was in a neighborhood I'd never been in, maybe west 76th street close to Broadway. Anyway, I walked into the veterinarian's office and didn't see anyone. I walked closer to the counter and noticed a small older woman who resembled Dr. Ruth Westheimer. "Can I help you?" she said in a thick

New York accent. "Yes," I answered, "My dog won't eat, and I'd like to see the vet." "We don't take walk-ins!" she screamed. "Well, do you want me to go outside and call to make an appointment?" "No, hold on a minute." She walked to the back and returned with an older gentleman with a lab coat on. "Follow me, sir," he said as he shuffled into a dark exam room.

As he walked in, he turned on the light which didn't do much. It looked like nobody had been in that exam room since about 1977. Old water-stained and faded pet photos littered the walls that sat above the sticky floors, which had a strong odor of stale dog kibble and medications. "What can I do for you?" he said as he turned around and put on his gloves. "She won't eat." I said. "Try this," he grumbled, and gave me a sample bag of food. He then turned to me quickly with nail trimmers. "She needs her nails trimmed!!" he yelled and grabbed her front paw, quickly cutting three nails really short. Greta screamed as blood gushed from her paws. "What are you doing?!" I asked. "She's fine," he said, dipping her paw in blot clot powder. "You're done." He walked out, and I sat there for fifteen minutes and saw nobody else. I walked out and nobody was at the desk, so I just left.

A bit traumatized by this scenario, I thought Greta needed some treats and definitely some new food. We walked back to the east side (at this point Greta was too big to take the train unnoticed). We went to this cool pet store on 96th and 2nd avenue. It was a small store but loaded up with all the goodies. I asked the clerk to show us their best pet food and he guided us toward the back. As Greta and I followed him, she stopped at a section that had these nasty looking treats. Feet, liver, jerky, whatever—if it was gross, it was in this aisle. She snatched something while the guy was showing me food. "She likes that huh?" he laughed. "Yeah, she does. What is it?" I asked. "Bull penis!" he laughed. "Okay, well put it in the bag with the food to choose and let's see what else she likes." We walked around the store and nothing else seemed to interest her but that nasty aisle of organ meats. She pulled me over to that aisle and picked those treats up again, so we took the bag of goodies home and that's all she ate. When I put the food in her bowl, she literally backed away like there was poison in it. I tried two other foods that were considered the best at the time, but she gave the same reaction.

On the way back to the pet store to return the bags of food, I put on my headphones and listened to the radio. I heard a report on dogs dying and food recalls, and the three bags I bought were on that laundry list of recalled foods. After I returned the food, I took Greta home and went to the grocery store to do some shopping. "What do wolves eat?" I said to myself as I filled the cart. I thought about the days as a kid when I'd study the encyclopedia for information on what made canines tick, what they ate, what breeds were out there and their places of origin. I thought about Heidi and the original Greta, and how bad food killed them. I didn't want her to die. I wanted to make food for her to live the best life she could. I walked through the store and got a whole raw chicken, carrots, broccoli, kale, apples, and Greek yogurt to make her first meal.

When I walked through the doors with the groceries, she seemed to zone in on the bag with her meal ingredients in it. I almost had to fight her off to get to the kitchen! Since this was my first time doing this, I was a bit clueless. *Should I cut up the chicken? What do I do with the vegetables? Is this going to kill her?* You know, normal things you think about when feeding a child. So, I decided the best thing to do was to shove the veggies in the chicken like it was a Thanksgiving turkey and top it with yogurt. Makes sense, right? Screw the plate—put that bird directly on the freshly cleaned wood floor and lock yourself in the bathroom until it's over. So into the bathroom I went, while Greta murdered the dead chicken and vegetables on the floor. *Crunch. Snap. Slurp.* Then, silence. I waited like eight minutes before coming out of the bathroom. I expected a mess, but I was shocked by what I saw: Greta sitting on the couch licking her lips, and the floor cleaner than it was before the chicken hit it. Nobody died, no shots fired...this was a success! But wait, what happened to those bones? She ate all those *bones*!!

The more I tried, the more she liked it. I'll admit, it was kind of gross hearing her crunch through bones and rip a bird apart, but this is apparently what would have gone down before humans. As far as I know, wolves didn't have thumbs to use silverware, so this is as good as it gets. She loved raw anything—gizzards, liver, feet, necks—whatever it was, she loved it. I started getting more creative with her meals, searching for ways to give her variety in her diet. I felt like since I didn't eat the same thing every day, why should she? I tried giving her steak, but that didn't go well. She used it as a chew toy and secretly stored it under a couch cushion and didn't tell anyone. I thought it was because she didn't like it, but I was wrong—she just didn't seem to know what to do with it. When I tried ground beef, however, I ground up fruits and veggies so she'd get them in her meal, too, without knowing. This was the trick that made the dish stick! She *loved* it! After finishing her dinner, she'd always go in the bedroom and people-watch while chewing on my basil plant.

I wasn't a native New Yorker—I was a DC transplant trying to survive as a musician and an actor taking all the jobs I could. I bounced from restaurants to film catering to make ends meet but it was still a struggle. After a while, living in New York became too much, so Greta and I moved to Baltimore.

It took a while, but I eventually got used to Baltimore. It was a bit slower—a *lot* slower—and the hustle was different. The people were a bit friendlier, and because it's a blue-collar city, it wasn't about being flashy. During the day, I would walk Greta through the streets of Baltimore to explore the city. She was quite the talking point, as it seemed there were no Dobermans in Baltimore. "Damn man, is that a Doberman? I thought they were extinct," I'd hear when we walked. While we were on a walk one day, several people approached me and asked what I fed Greta because she looked so healthy. "I make her food and treats myself," I'd say proudly. "How can I buy it?" they'd ask as I walked away. This happened for months, and then it occurred to me that I was onto something. I started to put the pieces together to make meals and develop a business plan.

At first, I'd go to doggy functions with my creations and people would laugh at me. "My dog would never eat that" they'd say, or "That's not dog food, that's people food!" While everyone had an opinion, the dogs' opinions of the food were paramount. Against the will of the humans, the dogs *loved* the food! I continued to do special tasting events at condominiums or doggy daycares and started booking clients for food deliveries. I felt I needed more proof, so I called veterinarians and started pitching. "Hello, my name is Kevyn Matthews, also known as The Dog Chef. Do you have any animals that you can't help?" is what I'd say to them. "Well, what do you mean?" they'd ask. I'd tell them I was a new doggy nutritionist who wanted to see if I could heal dogs with my food. I was surprised by the responses I got, both positive and negative. It was interesting that holistic veterinarians with track records of helping dogs permanently overcome huge obstacles embraced my philosophies, but classically trained veterinarians who only used man-made medications didn't want anything to do with me. While this was an obstacle, it didn't stop me from leaping over it. In fact, it made me build bridges to not only get over the obstacles, but lead veterinarians and their clients over the bridges to healthier, happier lives for their dogs.

I quickly became the talk of the town. "Who is this Dog Chef guy?" "Where did he come from?" Well, about six months after being in Baltimore, I got a call from the offices of *Baltimore* magazine. "Is this The Dog Chef" a man asked on the phone. "We think what you're doing is awesome and want to do a story about it!" "Uhhh, okay," I said nervously. "Great! We will see you in two weeks to do an interview!" I froze in my tracks for a moment...and then I screamed, "Yeah!!!"

It was happening. It was all starting, right here and now. I sat down and thought of the possibilities of this being a "thing." I thought of how cool it would be to go all the way: a pet store that was like a restaurant with fresh food and treats.

After the *Baltimore* magazine article, things started moving a bit faster. People were taking notice, but it was only the beginning. After being in Baltimore for about a year, I felt the need to move back to Washington, DC. I looked for apartments that would take dogs, which were very few that would take a Doberman. I finally found a place above Georgetown in a neighborhood called Glover

Park on Tunlaw Road, directly behind the Russian embassy. We loved the neighborhood. It had plenty of cool places to walk and visit, and I really felt at home in DC. It was my birthplace, where I lived before moving to New York City.

I got a job at a gym in DC, but barely slept while I continued to promote and talk about dog nutrition after work. Every weekend, I'd make appearances at local veterinary offices and dog functions to talk on the importance of healthy dog food and would get new clients. After a while, this grueling schedule caught up with me, so I decided to focus on my dream. In order to supplement my loss of income, I started making treat bags and selling them at events. I had carob chip cookies that looked like chocolate chip cookies, flax nut cookies, apple flax cookies, and kale pretzels. A few were updated recipes from when I was younger, and some were new. In the beginning, they didn't sell because they looked completely dangerous. They were all delicious—as you can imagine, with my loss of income, I sometimes had to eat a bag of dog treats for dinner.

As I was packing up to go to an event, the phone rang. "Hello," said a man over a broken connection. "I'm looking for The Dog Chef." "Yep, that's me, sir!" I replied. "I'm calling from Animal Planet's *Dogs 101* and would love to do a story on you. We think it would be great television." "*What?!*" I screamed and dropped the phone. When I picked it up, I heard his laughter and joined in. We talked for a while, and then he asked for my email address to send over the contract and filming dates. As he wrapped up, he said, "Don't worry about it Kev, it's gonna be fun. All television is bullshit, we're gonna make you a star!" and hung up. Holy crap. I mean seriously, I was in shock. I looked over at Greta, and it looked like she was smiling at me. "This is all because of you. I love you so much, Greta!!" I said as I gave her a hug. I reached into my box, dumped a whole bag of treats on the floor, and said, "Congratulations to us!" As I picked the box up and headed to the door, I turned to look at her and saw that she hadn't touched the treats. She was doing that stare again. "We did it," she said. "We did it." I walked out and stopped for a second. Part of me felt crazy that we had this connection where she could stare at me and I could hear her, but I wasn't afraid of it. I felt like she was guiding me to somewhere she needed me to take her, or us.

Over the next few months, Greta and I waited anxiously for our television debut. "Don't forget to check me out on Animal Planet," I'd say to random people, as if I was on more than one episode. When it aired, it was like a magic wand touched me and everyone had seen it. But for months, no sales came from it; at this point I was ready to quit. I walked to the mailbox one afternoon and was pleasantly greeted by an eviction notice. My landlord hadn't been paid in two months and was no longer cool with it. I walked back inside in shock and passed Greta without acknowledging her. The phone rang and I ran across the apartment to get it. "Hello?" I answered, slightly out of breath. It was my older brother, whom I hadn't spoken to in a while. I'm the youngest of three kids, so I idolized my brother. "Hey man, Mom told me about the television thing. I'm really proud of you!" he said. "I think you're onto something dude!" While I wanted to talk to him, I started to cry and didn't want him to hear it. "Let me call you back," I said abruptly. When I hung up, an avalanche of emotion came over me. I ripped up my mail and threw papers around, all while screaming and crying. I was happy to hear from him, but with no sales, an eviction notice, and a seeming end to my dream before it hit the TV screen, I'd reached my breaking point. As I sobbed and the papers flew through the air, I could see Greta. She hadn't moved since I rudely brushed by her without acknowledgment. The papers seemed to fall like rain in slow motion as she stood there looking at me. "I'm a *loser*," I cried as I fell to my knees. "I'll never be anything, and this is never going to work." She didn't move, she just stared that stare. "If you stop, they will die," she said. I stopped crying immediately and stood up. Her eyes didn't lose contact with mine, and as I wiped my face I heard, "We've got work to do." I felt something. Don't know if it was a push or a shove, but I felt it. The sadness in my heart seemed to melt away, and my chest filled with a sense of pride and faith. "You got it," I said. I packed up the belongings I knew were important enough to take with me, and from that moment on, she never, ever left my side. Not even to use the bathroom.

The next few years were bumpy as we bounced from place to place chasing the dream. I ended up moving back in with my mom for a bit until I could afford a new place, but toyed with the idea of moving back to New York. For as long as I can remember, Mom wasn't much of a dog person. She had dogs as a kid, but I think the experience of our childhood dogs being poorly housebroken ailed her. Greta was no Dachshund—she was a ninety-pound Doberman and somewhat menacing if you weren't a dog person, but Mom adored Greta. She was respectful of the furniture, she didn't pee on the floor, and Mom knew that if I left Greta in the house with her when I was gone, she was safe. Greta was getting a bit older now, but it was barely showing. She had a little gray hair on her muzzle, but still had that glide when she was up to speed on a walk. We enjoyed a smoothies every morning before our walk, and sometimes I'd make her salad around lunchtime as a snack. Yep, a Doggie Caesar salad. Sounds crazy, but old girl loved it.

As time went on, Greta didn't want to stay alone with anyone but me. If I left her at Mom's house too long, she would whine and keep her awake, and she would seem generally unhappy. "She don't want to be nowhere else but with you!" Mom would say. So I took her on my deliveries to New York, and she was such a trooper. I had a 1998 Volkswagen beetle with no heat, and on every trip, she sat in the passenger seat like my official helper. I felt like I was driving with her guidance.

The more we traveled to New York, the more notoriety we got. It was almost like we were famous, and to me, she was the star. We were on Animal Planet, major New York networks, local Baltimore and Washington news appearances, all to discuss animal nutrition. We were on fire, and all eyes were finally on us. The world was ready for this new, buzzworthy food and treat philosophy, and maybe Greta and I could have our own show. As soon as I thought about it, film producers started calling, each offer better than the last.

While this was exciting, I started to worry about Greta. On the trips to New York, she stopped sitting up the whole time, and would lie down instead. After a long day trip that started in the morning, we arrived at Mom's house. While I sat there, I got an email from Animal Planet to make a Halloween haunted house out of treats for their release party in Manhattan. I was excited to get this email and replied immediately that I could do it. As Mom and I sat, I looked at Greta. I noticed that her head was sitting low and her energy level wasn't the same. I picked up the leash and put it on her neck, but she didn't jump up as usual. I nudged her, and she jumped up but swayed a bit. "Is she okay?" Mom asked. "I don't know, she's never done that." I felt a bit of resistance on the leash, but pulled on it. "Come on old girl," I said as we walked through the door. As the sun set, we walked across the grass toward a bench where the dog owners would congregate. Normally, they'd bark at Greta and she'd go wild, but she didn't today. She walked slowly into a circle of dogs and bowed her head. The dogs walked toward her silently and started kissing her, and she them. "Is everything okay? I've never seen them act like this before," a woman said. "I don't know, ma'am. I don't think she feels well." When the dogs stopped kissing her, she backed out of the circle and her back legs gave out, but she immediately stood back up and turned to walk toward the house. As we slowly walked back, I looked down at her and she seemed completely different. The skin on her neck seemed to sag down and she looked a hundred years older. It felt like each step she took was an eternity. When we walked in, I noticed my mom looked terrified. "What's wrong with her?" she cried. "I don't know, Mom, but I need to get her to the veterinarian ASAP." At that moment, she started bobbing a bit. "She's not herself, Ma," I said as I touched her hip. When I added pressure to her hip, she fell over on her side. "Oh my God!" Mom exclaimed as Greta hit the floor. With tears in my eyes, I picked her up and put her on my bed. She was breathing and awake, but I knew I had to act fast. I ran to Mom's room to get the phone and call the veterinarian. "Hello, we have an emergency. I don't know what's wrong, but I need to bring her now." With the leash and collar in my hand, the veterinarian on the phone, and tears in my eyes, I ran to my room to put her collar on. As I reached to put it on, her head went limp. She was gone.

I had lost all motivation to do anything dog related, but I had already signed on to do that event, and my creation was to be the focal point of the evening. The event was in Manhattan on Halloween, and I was contracted to make a haunted house out of dog treats. With three days left before the event, I tried everything, but nothing made this project work. The dough failed and nothing stood correctly, so I trashed it. I called the promoter and told them what was happening, and they agreed to overnight a case of treats to finish the haunted house. The following day the treats arrived, but they were all broken. I panicked. I now had no plan at all to finish this project, and only one night to do it.

I tried to think of something scary, and the only house I thought of was the Bates Motel from *Psycho*. I'd only seen the movie twice as a kid, so I had to search for photos. With my tools in hand, I got to work. While working, I heard Greta's collar in the distance as it used to sound when she'd come running to me. "Greta?" I said, but no answer. As I sat back down, I felt something brush by my leg, and a sense of warmth took over the room. I knew it was her and I didn't want her to go, so I didn't say anything, I just started to create. At four in the morning, after working nonstop for eight hours, I was finished. I was exhausted and, so I went outside for a break. As I walked out the glass door, I noticed twelve deer, some standing, some sitting, but all with their eyes on me. One by one, they slowly turned around and walked away. When I came back inside, the warmth I'd felt while I was working all night had seemingly gone, but it was replaced with inexplicable confidence, faith, and direction. The following month, I was asked to do a Thanksgiving recipe for *Woman's Day* magazine, followed by a VIP invite to create goody bags for Beth Stern's Christmas Bulldog event, where I got a chance to meet Ice-T and personally give him my revolutionary CBD treats for his dog.

Life kept getting good for me, but something was definitely missing. I missed having a dog in my life but felt I could never replace Greta. After about a year, I got a call from the woman who gave me Greta. "Hey Kevyn, I know we haven't talked for a while, but I have a very special dog," she said. "She's from Greta's direct line and was going to be my puppy, but I can't keep her." Without hesitation, I said "I'll take her!" "You're gonna love her," she said. "She's something special." I drove to Ohio to get her and drove back with her in the backseat. "Your name is Heidi!" I said as she perked up and looked out the window. She rode quietly for the entire trip. I wanted my mom to see the new puppy, so I stopped by her house on the way back. "We're *here*!" I exclaimed as I pulled into the driveway. She yawned, stretched, and stepped out of the car, and as we walked to my Mom's door, I realized how relaxed she was. "Look at her!" Mom said with excitement. "She is absolutely gorgeous." As we walked through the door, I noticed that Heidi stopped, moved to the side of the door, sat down, and refused to move. She then laid down in the same spot with her head erect. When I went to move her, I noticed she was sitting on Greta's collar. I was speechless and overcome with emotion.

"She was your guardian angel, boy. She only stuck around to make sure you'd be okay. The moment she knew you were strong enough to move on, she moved on," my mother said, standing behind me. Then she smiled and said, "From the looks of this dog here, she's not done with you yet."

Chapter 2

The Essentials of a Dog Chef

When I started out making dog food, I had my mother's kitchen and all of her neat little tools at my disposal. In addition to those tools, I read all of her cookbooks and teaching guides, which helped me become a better chef. While you may not have all of those tools and gadgets, or an extensive knowledge of making dog food and treats, that's okay. I'm going to show you everything you need to create your own meals and treats!

If you're new to this whole cooking thing, there are a few things you'll need to get started. Most kitchens have the basic setup you need, but to be clear, you'll need a working oven, a sink with running water, a gas or electric stove that works, and a somewhat positive outlook on the outcome regardless of your experience in the kitchen thus far. This is for the dogs, so if it brings you anxiety while you're using it, chances are your dog may not like the food you're creating because it contains that energy. Relax. This may sometimes be a grueling task, but you are doing it for the love and life of your pet. Not saying you need to burn sage before you get started, just get excited. It's going to be fun, and your dog will love you for it. Let's go!

Tools

When I think of the tools you need to do this on a semi-regular basis, you're going to need these things:

* Food processor
* Chef's knife or knives
* Measuring cups
* Mixing spoons
* Mixing bowls (try to find a set with various sizes)
* Cupcake baking tins
* Cake decorating bag and tips
* Vacuum sealer and bags
* Measuring scale
* Parchment paper
* Cutting board
* Sheet pans
* Apron, or chef's coat if you wanna "look the part," LOL!

Food Processors

The first thing you'll need is a really good food processor. I've gone through a ton of these in my time before realizing that, if you spend the money on a good one, you'll only spend it once so spend accordingly. While they are more expensive, I would suggest buying a commercial one. They have longer warrantees, better blades, and are bulletproof. Since this is your primary piece of equipment, look to spend between one and five hundred dollars on a decent one that is also able to grind bones. I suggest buying a Robot Coupe processor. I have had mine for over ten years and it's still going strong. This, too, is a piece of cutlery, so please keep the blade sharp and never put your hands or anything else in a food processor while it's on. I know you aren't stupid enough to do it, but somebody is, and they may be reading this right now.

Chef's Knives

You don't need to spend millions on knives, just get sharp ones with a nice weight to them and make sure they feel good in your hands. Nothing will discourage you from learning something faster than using tools that you don't like. Try and get a set that includes knives of all sizes, as well as peelers, melon ballers, an apple corer, meat scissors, and a sharpener. Always keep your knives clean and sharp—dull blades are more difficult to use and are an easy setup for a nasty cut. Let's keep this fun and stay out of the ambulance, capisce?

Having said that, while we're talking about knife safety, here's a general crash course in proper knife technique and etiquette. The chef's knife is the most-used tool in any cook's arsenal. Mastering your chef's knife skills is the first thing they teach you in culinary arts school. The reason is because your speed and technique must be built from the ground up the proper way. If not, it can lead to bad habits that can result in injury, inefficiency, or other detriments. Since we're dog cheffing, speed is not so important.

When I first started taking cooking seriously, I had some bad habits with my knife skills, including improper knife handling, opposing hand position, and incorrect chopping motion. I had to relearn how to cut with a knife and it really felt like I was using my left hand. It takes at least six months to get a good feel for proper chef's knife techniques, but don't give up. You'll get it.

A chef's knife has a broad, tapered shape and a fine sharp edge. Its blade ranges in length from six to twelve inches, and measures at least one and a half inches at the widest point. It is designed to rocks on a cutting board as it cuts food.

How to Hold a Chef's Knife

For the non-knife hand:

* Your fingers are curled under to protect your fingertips.
* Your thumb and little finger are behind your other fingers.
* The side of the blade (but not the edge) rests against the middle knuckles of your **non-knife** hand. This helps you keep the knife from coming down on your fingers. It also helps you measure the size of the cuts as you move your hand backward on the food after each cut in preparation for the next one.
* Cuts are made downward with a rocking motion from the tip to the end of the blade. The knife is not sawed back and forth through foods.

For the knife hand:

* View how the thumb and first finger grip the blade just beyond the handle. This helps make the knife an extension of your arm and gives you better control and precision when cutting.
* The tip of the blade is kept in contact with the cutting board. The blade is rocked up and down until the food is chopped to the desired size.
* To prevent vegetables and fruits from slipping on your cutting board, cut them in half before slicing or chopping further. This helps anchor them firmly on your cutting board and helps protect against cutting yourself.

What Size Chef's Knife Do You Need?

Get a good knife set with several sizes to fit your needs. Using a knife that's too small is dangerous if you're trying to cut something large because the knife can slip and go through your hand.

Things to consider when buying a knife set:

* Do they feel comfortable when you hold them in your hand?
* Can you easily manage them when you go through the motions of slicing and chopping?
* Do the blades feel solid, not lightweight and flimsy?

Caring for a Chef's Knife

Below are some tips to help prolong the life of your chef's knife:

* Many knife companies recommend not washing your knife in the dishwasher to avoid damaging the blade. Also, wooden handles may not hold up well when washed in a dishwasher. Always dry the knife before storage.
* Follow the manufacturer's directions for sharpening your knife. A sharp knife not only cuts better but is safer than a dull knife. There's a tendency with dull knives to use too much force, lose control, and cut yourself.
* Avoid cutting on hard surfaces that dull the edge of your knife, such as glass cutting boards. Softer cutting boards, such as polyethylene plastic cutting boards, are much easier on knives.
* Store your knives in some type of knife block or other storage system that keeps the blades separate. Do not throw them together in a drawer where they can bump against each other and possibly damage or dull their blades.
* Always make sure the cutting board is secured to the counter with a wet cloth or paper towel. It's dangerous to have a cutting board that moves around while you're trying to cut.

Measuring Cups and Spoons

One of the most important things in baking and making dog food is being consistent. While measuring cups are important and an instrumental part in everything you do here, they don't need to be fancy. Measuring cups and spoons come in different materials. I prefer plastic over metal, because metal can dent and alter the measuring capacity. Plastic versions of both of these can be found at any grocery or baking supply store.

Mixing Spoons

Mixing spoons are not all created equal! They come in several materials, shapes, sizes, and colors. I generally go for classic wood mixing spoons due to their strength and durability. However, if they sit in moisture too long, they become brittle and splinter, so they require a bit more care. Plastic mixing spoons are also okay and easier to clean than wood spoons, but they're sometimes flexible when on the cheaper end of the price scale. Metal spoons like soup spoons are good and sturdy for mixing, but they are sometimes too short. Larger metal mixing spoons are definitely a no-no for me, as they bend too easily.

Mixing Bowls

I have found that the best type of mixing bowls are chef-grade steel bowls. You can get a small to large set of these bowls. These are commonplace in every commercial kitchen and having a complete set will allow you to use them in a variety of applications. Steel bowls are easy to clean and indestructible. You'll need the larger bowl to create meals for sure.

Cupcake Baking Tins

For any of the muffins or cupcakes in this book, you'll need cupcake tins. These are also available in many different material options. They come in metal and silicone, but I prefer the old-school metal kind. They're easily accessible at most grocery stores and they work the best. You can find different sizes, as they have mini, regular and jumbo size. If you aren't sure, get all the sizes so you can be flexible when creating. These can also be used to make candy and other cool treats!

Cake Decorating Bag and Tips

In order to make awesome cake designs, you'll need a decorating bag, tips, and a quick crash course in piping design. I like to use disposable piping bags because there's no cleanup, but different brands offer a variety of materials, like canvas and vinyl. Canvas bags don't work well for the icing you'll be making in this book. Vinyl bags, however, are great and can be reused for quite a while before needing to be replaced. Decorating tips can be plastic or metal. While metal ones are more precise, plastic ones are more durable and won't rust. Try to find a piping tip set that has all the tips so you can be as creative as possible when making treats.

Chapter 3

Digestion and the Immune System

I originally started making dog food because of my fascination with their digestive systems. As humans, we seem to think that every digestive system is like ours, which is not the case. Before we eat, we prepare our food by taking certain steps. Usually, we wash our food to get rid of any toxins on the outside before cooking it. If we're cooking fruits or vegetables, the heat begins the process of breaking the food down for digestion. Humans usually cook their meats to kill off any disease or bacteria that we cannot break down in our digestive tract, in addition to seasoning the meat to make it palatable.

Once chewed, food is pushed by the tongue into the esophagus and swallowed. Through a process called peristalsis that uses the muscles in the walls of the digestive system to move and force food downward, the food then heads to the stomach and GI tract for further breakdown. Once food leaves the stomach, it empties itself into the small intestine as a substance called chyme. The small intestine has three parts: the first part is called the duodenum, the second (middle) is the jejunum, and the ileum is at the end where it meets the large intestine. The small intestine gets help from the pancreas and the liver to make digestive fluids that further aid the process, while the walls of the small intestine absorb the water content and the nutrients from the food and releases them into the bloodstream. At this point, your food moves to the cecum, which is the first part of the large intestine, then to the colon, and finally to the rectum, and well, you know what happens next...*poop!*

This process in humans, from start to finish, can take anywhere from twenty-four to seventy-two hours depending on the food you eat and your lifestyle. While food seems to travel quickly in the beginning of the process (which takes six to eight hours), the rest depends on factors like gender, metabolism, and digestive issues, if any. Dogs, however, are a completely different animal (no pun intended).

A key element to the way people eat is preparation of our food for breakdown *before* we eat it. Digestion starts in the mouth, as our teeth break food into smaller pieces and grind it up and send it to the stomach through the esophagus. Water in saliva helps turn food into a soft paste that is easier to swallow. Our saliva also contains chemicals called enzymes which break the food down into smaller molecules and further aid digestion. With dogs, digestion starts as the mouth bites and pulls food down into the esophagus. While dogs produce saliva for moisture to help the food move through the esophagus, their saliva contains a special enzyme to aid any injury to the mouth during the takedown of food. Dogs know the healing effect of their saliva, which is why they lick any cuts they get and even try to heal a burn or cut we may get to make us feel better (even if it doesn't). How cool is that?!

Have you ever noticed a dog's mouth while chewing only can make two motions with their jaw? Up, down...*right*? Now, you (human) go eat a piece of chewy candy, and what happens? What happens when a cow chews grass? Or a horse chews a carrot? Well, a human jaw (like a cow's or horse's) can pivot because we were meant to *chew*!! What do a wolf, Chihuahua, snake, and alligator have in common? Their jaws aren't designed to chew side to side. They're designed to bite, pull, and swallow. In fact, snakes dislocate their jaw every time they eat. Does this mean your Chihuahua will eat your kids? I hope not.

The next step in their digestion is similar to ours because they also use their esophagus to move food downward to the stomach. However, human stomachs are a bit different from this complicated machine. A dog's stomach is a hundred times stronger than ours, with more powerful acid to break down the proteins and activate the enzyme pepsinogen to release amino acids into the bloodstream. The acid is also capable of softening bone matter, which is how wild dogs are able to eat and digest entire animals without getting sick. After the stomach, food is sent to the small intestine, where your dog's body absorbs the nutrients that they need from their food. After the nutrients are absorbed, the remaining waste products follow the path to the larger intestine and leave the body. The entire process for a dog to digest food can be anywhere from four to nine hours, depending on the type of food they're fed. If you decide to feed them raw or fresh food, the window for digestion is four to six hours. If you feed them store-bought kibble, it's nine hours (sometimes more) for digestion. Big difference, huh?

Let's talk about this for a minute...

A dog's intestinal tract is roughly two feet, whereas a human's is twenty feet long—half the length of a badminton court. This would explain why your dog sometimes poops out foreign objects that look like they didn't process through the system at all. If the system doesn't recognize it, it's immediately released from the body.

When feeding fresh food, the canine body is more likely to stay in tune with its immune system's natural functions. Dogs weren't designed to eat processed food—the lack of live "good bacteria" in this type of food causes it not to break down as quickly and spend more time in the digestive tract. Dry processed foods can sometimes cause inflammation due to the lack of water, which leaves no way for nutrients and moisture in the food to be used within the system which can cause issues down the line. A healthy canine digestive system has the right balance of "beneficial" bacteria and correctly functioning immune cells, and its health is maintained by a diet that is appropriate for the individual dog. What a dog eats has an impact on their entire body, including their digestive system. A diet catered to your dog's needs can go a long way to managing your dog's digestive and immune systems.

A dog's immune system consists of a network of white blood cells, antibodies, and other substances that fight off infections and reject foreign proteins. In addition to that, the immune system includes several organs. The thymus gland and bone marrow are the sites where white blood cells are produced. The spleen and lymph nodes trap microorganisms and foreign substances and provide a place for immune system cells to collect, interact with each other and with foreign substances, and generate an immune response.[1] This is super important, because dogs are under constant threat of microbial invasion. These potential invaders gain access to the body through the intestine and respiratory tract and the skin. "Good bacteria" in the intestine serve to protect your dog from infectious invaders by occupying a niche that precludes other organisms from establishment there.

To prevent microbial invasion, the body has (as part of the innate immune system) a series of defenses that collectively constitute a highly effective defense against invasion. These mechanisms include the skin, which has its own microbiota and utilizes desiccation as a mechanism to discourage colonization with other organisms. Inhaled microorganisms and other material are rapidly removed by the mucociliary apparatus, which consists of ciliated epithelial cells and mucus-secreting cells that move inhaled material from the lower to the upper respiratory tract from which they are removed by the cough reflex.

The second line of defense is "hard-wired" and depends on a rapid stereotypical response to stop and kill both bacteria and viruses. This is typified by the process of acute inflammation and by the classic illness responses such as a fever.

The third line of defense is the highly complex, specific, and long-lasting adaptive immunity. Because an animal accumulates memory cells after exposure to pathogens, adaptive immunity provides an opportunity for the host to respond to exposure by creating a highly specific and effective response to each individual infectious agent. In the absence of a functional adaptive immune system, survival is unlikely.

---

1 https://www.merckvetmanual.com/immune-system/the-biology-of-the-immune-system/biology-of-the-immune-system-in-animals

Food[2] that contains prebiotics can be effective in helping rebalance your dog's microflora and contribute to a healthier digestive tract. I know, I know... You're probably thinking, *What are prebiotics? I thought my dog needed probiotics!!* Well, prebiotics are a type of fiber that the body cannot digest. They serve as food for probiotics, which are tiny living microorganisms (the "good bacteria"). Both prebiotics and probiotics support helpful bacteria and other organisms in the gut. Highly digestible and high-quality proteins are essential in providing your dog with the building blocks for cell growth in their body. Food that uses proteins from high-quality sources can help maintain a healthy digestive system and help avoid making their digestive sensitivities worse.

Dogs also need the right balance of fiber in their diet. A high-quality food should have just the right balance of fiber to make sure your dog can absorb nutrients easily without causing unnecessary strain on their digestive and immune systems.

2 https://www.royalcanin.com/vn/vi/dogs/health-and-wellbeing/what-makes-your-dogs-digestive-system-healthy

Chapter 4

The Meat and Potatoes: What Dogs Can and Can't Eat

Exploring a dog's palate can be an interesting thing, since they don't enjoy food as we do. For the most part, 80 percent of a dog's meal is in his nose. What I mean is that a dog decides to eat something because of the smell, not necessarily the taste. With only 1,700 taste buds, their tongues only have four signals to send to the brain to tell it whether something is salty, sweet, hot, or cold. So in essence, the dog kind of does a pairing of the senses before eating.

When discovering what foods your dog likes, there are definitely things to consider. If you have any worries about your dog's potential allergies going into this chapter, you may want to check with your veterinarian about which foods your dog needs to stay away from. Whether you're making any of the recipes in this book or creating your own doggie meals and treats, here's a great guide on what you can and cannot use.

Household Things Your Dog Should Never Consume[3]

* **Antifreeze** contains ethylene glycol, a sweet-tasting, odorless substance which can also be found in lower amounts in several windshield de-icing agents, hydraulic brake fluid, motor oils, solvents, paints, film processing solutions, wood stains, inks, and printer cartridges. Dogs are attracted to ethylene glycol by its sweet smell and taste and will voluntarily drink it if it is spilled or leaked onto garage floors or driveways. Ethylene glycol has a very narrow margin of safety, which means as little as half a teaspoon per pound of a dog's body weight can result in fatality. Within thirty minutes of ingestion, signs of lethargy, vomiting, incoordination, excessive urination, excessive thirst, hypothermia (low body temperature), seizures, and coma can occur twelve to twenty-four hours after ingestion. Some of the signs seem to dramatically improve, luring pet owners into a false sense of security. However, during this stage, dogs become dehydrated and develop an escalated breathing and heart rate. After thirty-six to seventy-two hours, signs of severe kidney dysfunction, which is characterized by swollen, painful kidneys and the production of minimal to no urine, may occur, as well as progressive depression, lethargy, lack of appetite, vomiting, seizures, coma, and then death. It is critical that you bring your dog to a veterinary clinic if you know or even suspect that they have consumed antifreeze, or if they are exhibiting any of the early symptoms. Do not wait; time is of the essence and immediate treatment is essential! Dogs must be treated within eight to twelve hours of ingesting antifreeze, as the antidote only has a narrow time period to work. Left untreated, your dog will die.

3 https://vcahospitals.com/know-your-pet/ethylene-glycol-poisoning-in-dogs

* **Chocolate** is derived from the roasted seeds of *Theobroma cacao*. The primary toxic principles in chocolate are the methylxanthines theobromine and caffeine. Although the concentration of theobromine in chocolate is three to ten times that of caffeine, both constituents contribute to the clinical syndrome seen in chocolate toxicosis. The exact amount of methylxanthines in chocolate varies because of the natural variation of cocoa beans and variation within brands of chocolate products. Signs of chocolate toxicosis usually occur within six to twelve hours of ingestion. Initial signs may include polydipsia or abnormally great thirst, vomiting, diarrhea, abdominal distention, and restlessness. Signs may progress to hyperactivity, polyuria, ataxia, rigidity, tremors, and seizures. Tachycardia, premature ventricular contractions, tachypnea, cyanosis, hypertension, hyperthermia, bradycardia, hypotension, or coma may occur. Hypokalemia may occur later in the course of the toxicosis, contributing to cardiac dysfunction. Death is generally due to cardiac arrhythmias, hyperthermia, or respiratory failure. The high fat content of chocolate products may trigger pancreatitis in susceptible animals. One ounce of milk chocolate per pound of body weight is potentially lethal in dogs.
* **Macadamia** nuts are cultivated from the continental USA, Hawaii, and Australia. Their level of toxicity is not known. Dogs have shown signs after ingesting 2.4 grams of nuts per kilogram body weight. Within twelve hours of ingestion, dogs develop weakness, depression, vomiting, ataxia, tremors, and/or hyperthermia. Tremors may be secondary to muscle weakness. Signs generally resolve within twelve to forty-eight hours.
* **Avocado**[4] contains persin, which is a fungicidal toxin, that can cause serious health problems and even death according to some veterinarians. While dogs seem to be more resistant to persin than other animals, that doesn't mean avocados are 100 percent safe for your dog to consume. Persin is present

4 https://www.akc.org/expert-advice/nutrition/can-dogs-eat-avocado

in the avocado fruit, pits, leaves, and the actual plant, so all of these parts are potentially poisonous to your dog. Most of the persin is concentrated in the leaves, skin, and pit of the fruit. It is also present in avocado flesh in small amounts. Exactly what amount of persin is lethal isn't known. In large amounts, it can cause vomiting, diarrhea, and myocardial damage. Avocado flesh's high fat content can lead to gastrointestinal upset and pancreatitis in dogs if they eat too much, and because it's calorie-dense, it can also lead to weight gain. Another concern is the pit at the center of the fruit, which may cause choking. Strains of avocado come from all over, and some contain less persin than others. While some have been used in some dog products, it's best to leave them alone.

* **Xylitol**[5] is a naturally occurring substance that is widely used as a sugar substitute. Chemically, it is a sugar alcohol, and it is found naturally in berries, plums, corn, oats, mushrooms, lettuce, trees, and some other fruits. Commercially, most xylitol is extracted from corn fiber, birch trees, hardwood trees, and other vegetable material. Although it has been used as a sugar substitute for decades, its popularity has increased dramatically in the last few years due to its low glycemic index and dental plaque-fighting properties for humans. When a dog eats something containing xylitol, it is quickly absorbed into their bloodstream, resulting in a potent release of insulin from the pancreas. This rapid release of insulin causes a rapid decrease in blood sugar levels (hypoglycemia), an effect that occurs within ten to sixty minutes of eating the xylitol. Untreated, this hypoglycemia can be life-threatening. The higher the dose ingested, the more the risk of liver failure. The most common source of xylitol poisoning comes from sugar-free gum, which would take up to nine pieces of gum to result in severe hypoglycemia in a forty-five-pound dog, while forty-five pieces would need to be ingested to result in liver failure. As there is a large range

5 https://vcahospitals.com/know-your-pet/xylitol-toxicity-in-dogs

of xylitol in each different brand and flavor of gum and other products, it is important to identify whether a toxic amount has been ingested. Either way, call your vet immediately if you see your dog consume this one.

* **Alcohol**[6] can be found in some surprising places. Rum-soaked cakes, yeast dough, or unbaked deserts containing alcohol may cause poisoning in pets just as much as a cocktail. While it might seem harmless to let your dog take the tiniest sip of your wine, beer or mixed drink, the bottom line is that it's never okay to let your dog drink alcohol. It's never acceptable to put their health at risk, no matter how amusing it may seem at the moment. As a pet parent, it's your responsibility to keep your pooch safe, and that includes keeping him away from alcohol. Dogs who have consumed toxic amounts of alcohol will begin to show effects within thirty to sixty minutes. Symptoms can range from mild intoxication to severe inebriation that can be life-threatening and may show the following signs of alcohol poisoning. Without a doubt, call the vet.

    * Vomiting
    * Disorientation
    * Inebriation
    * Loss of bodily control
    * Diarrhea
    * Hyper salivation
    * Excitement that changes to depression
    * Difficulty breathing
    * Loss of consciousness
    * Dehydration
    * Slow heart rate
    * Seizures
    * Heart rhythm problems[7]

6 https://www.hillspet.com/dog-care/healthcare/what-happens-when-dog-drinks-alcohol

7 https://wagwalking.com/condition/alcohol-poisoning

* **Grapes and raisins**[8] may seem harmless, but they're not to dogs. Even small amounts of grapes or raisins can be fatal. The issue is that not all raisin and grapes will exert similar results. Moreover, not all dogs will react to the toxic principles of raisins and grapes. One of the most serious complications of grape/raisin toxicity is sudden kidney failure with lack of urine production. Vomiting and diarrhea often occur within the first few hours, and twenty-four hours after ingestion, vomit and fecal contents may contain pieces of grapes and/or raisins. Other symptoms include abdominal pain, dehydration, and unusual quietness and weakness. This is an emergency which requires immediate treatment. If your dog ingests grapes or raisins, you will need to induce vomiting as soon as possible before the toxins in the fruit can be absorbed. Contact your veterinarian immediately.
* **Yeast**[9] can rise and cause gas to accumulate in your dog's digestive system. This is painful and can cause the stomach to bloat and potentially twist, causing a life-threatening emergency. Because the yeast produces ethanol as a by-product, a dog that ingests raw bread dough can become drunk, and the effects of alcohol consumption will follow. (See above.)
* **Caffeine** may be a great way to start your day, but most people may not realize that caffeine can be harmful to their pets. You also may not know that many foods and drinks in your cupboard contain caffeine. Most people name coffee as the number one source of caffeine, and they are right. Most households have coffee in the pantry. However, tea and soda are full of caffeine, as well as those energy or sports drinks in the fridge. Other common sources of caffeine are diet pills and over-the-counter products that keep you up all night. These products are loaded with caffeine, and deadly to your dog.

8 https://www.petsmagazine.com.sg/a/features/pet-pantry/254/grape-and-raisin-toxicity-in-dogs

9 https://www.aspca.org/pet-care/animal-poison-control/people-foods-avoid-feeding-your-pets

How to Induce Vomiting if Chocolate or Any Other Toxin Is Consumed[10]

You will need:

* 3 percent hydrogen peroxide. ***Only use 3 percent!***
* 1 teaspoon
* Paper towels

1. Administer one teaspoon of hydrogen peroxide (5 ml) per ten pounds of body weight.
2. You may administer this dose a maximum of two times. Therefore, administer the hydrogen peroxide dosage once and then wait fifteen to twenty minutes. Walking your dog around may help expedite the process.
3. If the dog does not vomit, repeat the dose. If they still do not vomit after another ten minutes, bring your pet and the chemical bottle or other toxin to the vet at once, as the vet may have more effective products to induce vomiting.

Note: The bottle of hydrogen peroxide may say "toxic to pets." This simply means that it makes them vomit. If you follow the instructions and dosages carefully, your dog will vomit and there will be no long-lasting effects from ingesting the hydrogen peroxide. This needs to be done within twenty minutes of ingestion of the toxin.

10 https://pethelpful.com/dogs/How-and-when-to-induce-vomiting-in-your-dog

Plants Your Dog Shouldn't Consume (Indoor and Outdoor)

* **Poinsettia** toxicity comes from the milky white saps that are found on the plant, known as diterpenoid euphorbol esters. So yes, poinsettias are toxic, but it must be noted that the levels of toxicity found in poinsettias are mild.
* **Amaryllis** causes dogs to salivate excessively and experience abdominal pain. They may vomit and/or have diarrhea.
* **Asparagus fern** berries when eaten can cause gastrointestinal distress, with symptoms of vomiting and diarrhea, and the sap can induce a contact rash.
* **Cyclamen** (or sowbread) causes vomiting and diarrhea, and in extreme cases can lead to seizures and even death.
* **English ivy** can be harmful if ingested. It contains saponins, which cause abdominal pain, excess salivation, diarrhea, and vomiting. Other plants that contain saponins include:

    * **Aloe** leaves and roots (using the topical gel is fine)
    * **American Holly** (also known as inkberry and winterberry)
    * **Hosta**

* **Elephant ear** causes alocasia poisoning, which results in pawing at the face and mouth, drooling, foaming at the mouth, and vomiting.
* **Pothos**[11] produces insoluble calcium oxalate crystals. The crystal shape of the oxalates and their insolubility cause damage to the mouth. Instead of dissolving when coming into contact with the moisture of the mouth, they cut the

11 https://wagwalking.com/condition/pothos-poisoning

tissue and cause injury. Other plants that contain calcium oxalates include:

  * **Begonia** roots. These have the highest level of calcium oxalate, and cause vomiting and diarrhea.
  * **Lily**, which causes irritation, drooling, vomiting, and difficulty swallowing.

* **Jade** is part of the Crassulaceae family, and all the plants in this family are poisonous to dogs. Consuming this plant causes vomiting, lethargy, abdominal pain, weakness, and depression. In extreme cases, it can lead to a slowed heart rate and convulsions.
* **Azaleas** are highly toxic and eating a few can cause an immediate reaction. Azalea irritates the mouth and causes vomiting and diarrhea. Eating its leaves can kill your dog, so contact poison control or your vet immediately to find out the steps you need to take if your pooch gets into this plant.
* **Bay laurel,** also known as bay leaf, is a popular garden plant and seasoning in my hometown in Maryland. It contains eugenol, which is toxic to dogs. Bay leaves can cause excess salivation, vomiting, and kidney failure.
* **Cherry tree** leaves and branches contain cyanogenic glycosides, which are harmful to many species, including humans. The cherry pit contains cyanide. If your dog eats cherry tree parts, they may exhibit breathing difficulty, dilated pupils, and go into shock.
* **Wisteria** flowers and seeds are toxic. Dogs who eat wisteria may be confused, dizzy, nauseated, and have stomach pains and diarrhea. They may also collapse or vomit repeatedly. If they consume enough, it can even cause death.
* **Sago palm** is one of the most toxic plants to dogs. It causes diarrhea and bloody vomiting. If enough is ingested, it can cause bleeding disorders, liver failure, and death.
* **Pacific yew** leaves can cause tremors, respiratory distress, vomiting, seizures, and heart failure which may lead to death.

* **Oleander** is popular in the south and is considered one of the most toxic plants to dogs. It contains a substance called cardiac glycoside, which can cause heart irregularities, muscle tremors, vomiting, and diarrhea.
* **Nicotine** can cause hyper excitability, which is then followed by depression, vomiting, lack of coordination, paralysis, and even death.
* **Morning glory** can cause issues if your dog consumes a relatively large amount. It's moderately poisonous and can cause vomiting, diarrhea, and agitation or incoordination.
* **Marijuana** is said to be poisonous to dogs, so it's important to note the difference between marijuana and CBD. Many people advocate for using CBD oil as a treatment for dogs. The oil has a different composition and won't harm your dog. When consumed in larger amounts, marijuana can cause depression, incoordination, excess salivation, and dilated pupils. In severe cases, it can put your dog in a coma.
* **Locust** bark, leaves, and seeds are all toxic to your dog. They can cause vomiting, bloody diarrhea, and difficulty breathing, which can lead to death.
* **Mistletoe** of the American and European varieties are poisonous in large enough quantities. The plant can cause drooling, vomiting, diarrhea, abdominal pain, ataxia, seizures, and even death.
* **Juniper** is mildly poisonous to dogs. The berries, needles, and stems contain compounds that can make your dog sick with vomiting or diarrhea.
* **Hops** can cause panting and an increase in body temperature, which can lead to seizures and even death. Furthermore, the yeast in beer is also bad for your dog.
* **Fleabane** grows wild. It has a reputation for keeping fleas away and is used as a tea for urinary tract infections. It's a mild toxin for dogs that can cause stomach upset and indigestion.

* **Dieffenbachia** causes stinging in the mouth, excessive drooling, and swelling in the mouth and throat.
* **Daffodils** cause vomiting, diarrhea, and abdominal pain, which may lead to a drop in blood pressure and seizing. Tulips have these same effects.
* **Daisies** contain sesquiterpene, which can cause excessive salivation, vomiting, diarrhea, and uncoordinated movement.
* **Foxglove** has life-saving properties that can help people with heart failure. However, if consumed by a dog, it can make their heartbeat slow or irregular.

Fruits and Veggies Your Dog Can and Should Consume!

* **Apples**[12] are an excellent source of vitamins A and C, as well as fiber for your dog. They are low in protein and fat, which make them the perfect snack for young and senior dogs. Beware of the seeds, as they contain a small amount of cyanide and should not be consumed.
* **Bananas** are high in potassium, vitamins C and B6, biotin, fiber, and copper, and low in cholesterol and sodium. But due to their natural sugar content, you should feed them to a dog in moderation or as a treat, not as part of a daily meal.
* **Blueberries** are a low-calorie superfruit brimming with vitamin A, minerals, fiber, and some of the highest amounts of antioxidants in any fruit. They contain phytochemicals and anthocyanins, which fight cancer and reduce inflammation.
* **Broccoli** isn't just loaded with vitamin K, it also contains vitamins A, C, D, E and B 6, 9, and 12. Vitamin K is essential in building and maintaining strong bones in both young and older dogs. While other vegetables may contain vitamin

12 https://www.akc.org/expert-advice/nutrition/fruits-vegetables-dogs-can-and-cant-eat

K, none have this high a percentage. It also contains fiber, chromium, potassium, and manganese, which help grow muscle, lower blood sugar levels, enhance cardiovascular health, lower cholesterol, and reduce the risk of cancer.

* **Brussel sprouts**[13] are high in fiber and antioxidants, which reduce inflammation in the body and improve overall blood circulation. They're also loaded with vitamins C and K, but not as much as broccoli. Brussel sprouts are great for a dog's immune system and bone health.
* **Cantaloupe** is great for overweight dogs as a healthy snack to aid in weight loss. This delicious melon is loaded with vitamin A (as beta carotene), which helps reduce the risk of cancer and prevent cell damage. It's also a good source of vitamins B6 and C, fiber, folate, niacin, and potassium. In addition to all this, it's good for the eyesight and great for dogs with kidney disease. Avoid feeding your dog the rind of the cantaloupe though, as it can cause intestinal damage.
* **Carrots** have nutrient-dense properties. Whether raw or cooked, carrots can be a healthy addition to your dog's diet. Every part of the carrot can be good for dogs, including the leafy greens at the top. Carrots are a good source of fiber, vitamins A and K, potassium, and beta carotene. However, carrots should not be used as a full meal substitute to your dog's primary food. Carrots contain high amounts of natural sugar, so while they are safe for dogs to eat, they should be enjoyed in moderation. Too high of a sugar intake could cause weight gain and other health issues later in your dog's life.
* **Cranberries**[14] are rich with antioxidants and nutrients that help support your dog's immune system and decrease inflammation. Antioxidants play a major role in keeping dogs healthy. They're known for improving cognitive function and alleviating allergies and skin problems. Cranberries are low in calories and high in vitamin C, fiber, and potassium.

---

13 https://www.rover.com/blog/can-dogs-eat-brussel-sprouts

14 https://www.petmd.com/dog/nutrition/healthy-foods-checklist-cranberry-dogs

Cranberries also lower the risk for developing heart disease, stroke, hypertension, diabetes, and certain gastrointestinal diseases. Among other benefits, cranberries can improve your dog's bladder health, reduce tartar and plaque buildup, fight bacteria, and help prevent cancer. You can feed raw or cooked or cranberries to your dog in moderation. Avoid cranberry sauce and cranberry juice, which are high in sugar and may contain other ingredients that are potentially harmful to dogs. Feeding large amounts of cranberries to dogs can lead to the development of calcium oxalate stones in the bladder.

* **Cucumber** is a great snack for those with an overweight dog trying to shed a few pounds. Its low-calorie count and high-water content are a stark contrast from store-bought treats, so if you just got in from a long walk with your dog, it's a great way to quickly hydrate them. Never feed a dog a whole cucumber, as there is a risk of choking, and too much cucumber can cause issues in their GI tract. Cut it into small pieces to enjoy in moderation. For the record, while pickles are cucumbers, they contain spices and dangerous sodium levels that should *not* be consumed by dogs.
* **Green beans** contain vitamins A, C, and K, as well as other nutrients like iron, magnesium, and potassium. Green beans can, therefore, safely be given to dogs. Many of the nutrients found in green beans should already be covered in a dog's regular diet, so if you use green beans to create a meal for your dog, only use fresh or frozen. Never use canned vegetables, as they contain too much sodium for dogs.
* **Kale,**[15] like many other leafy greens, is very nutrient-rich. Cruciferous vegetables are known to be "nutrition powerhouses," and the potential benefits are many. Kale is loaded with vitamins A, C, and K, and contributes to higher energy levels, blood and muscle health, and a better immune system. It comes with the potential to fight cancer as well as other

15 https://rawbistro.com/blogs/raw-bistro/can-dogs-eat-kale

inflammatory diseases. It's a low-calorie treat that not only tastes good but can also aid with digestion. Kale is a known antioxidant, and is also high in calcium, magnesium, potassium, and iron. Kale can also support vision and colon health, aid liver detoxification, and help fight off infections. Feeding too much kale can lead to nutrient deficiencies—specifically amino acids, as it's low in protein. Therefore, kale should always be fed in moderation alongside a source of protein. When fed in this way, it can help minimize supplements needed.

* **Mango** is high in fiber, as well as vitamin A, B6, C, and E, making it quite nutritious for both humans and dogs. The snack is also sweet, so your dog will probably love it, but don't go overboard. The fruit is soft when ripe, but you should still cut it into small pieces, and of course, don't feed the skin or pit to avoid choking.
* **Oranges** are full of nutrients, potassium, and fiber. They are low in sodium, which makes them a healthy snack if given in limited amounts. This fruit is also full of vitamin C, which can benefit your dog's immune system. While some dogs may not enjoy the acidic taste of this citrus fruit, it's definitely good for them. However, if your dog has diabetes or any other health condition, the natural sugar content in oranges could cause an issue if you overfeed them. Use these as a snack or an ingredient in a creation, but not as a meal. Oh, and be sure to remove the skin and seeds!!
* **Peaches**[16] contain vitamins A and C. They are low in calories and high in fiber. As a source of antioxidants, they can help ward off cancer and boost the immune system. They also help improve the functions of the liver and kidneys. If served properly and in moderation, peaches are healthy snacks that can even be used as a refreshing reward during training sessions. Cut fresh peaches into pieces or use frozen peaches, but under no circumstances should you feed your dog canned

16 https://dogtime.com/dog-health/dog-food-dog-nutrition/59103-can-dogs-eat-peaches

peaches. They have too much sugar and other chemicals that make them unhealthy. When using fresh peaches, remove the pit and don't feed it to your dog. Like avocado pits, the pit or "stone" of this sweet summery fruit is a choking hazard.

* **Pears** help lower blood pressure and cholesterol levels, which are fundamental in maintaining heart health. Their high fiber content plays an important role in preventing cancer in the colon, lungs, and stomach. Pears can help prevent macular degradation linked to aging in addition to maintaining blood glucose levels. Pears are low in calories, rich in nutrients, and incredibly hydrating, which aids in avoiding heatstroke.
* **Peas** can help boost immune system function and reduce symptoms of arthritis, as they contain anti-inflammatory properties. They are loaded with antioxidants like alpha and beta carotene, which may slow down your dog's aging process. The fiber in peas can help prevent constipation and improve bowel health. Peas contain vitamins A, C, and K, phosphorus, potassium, and polyphenol, which has anti-cancer properties.
* **Pineapple**[17] contains bromelain, an enzyme that's also present in meat tenderizer. Rumor has it that bromelain makes poop taste foul, making it less appealing to those dogs that consider it a delicacy. I haven't used it for that, so I don't know if this is true. However, I do know that pineapples are 83 percent water—definitely great for hydration—and do contain vitamins A, C, and B6. They're also a good source of thiamine, riboflavin, magnesium, manganese, potassium, beta carotene, and antioxidants.
* **Raspberries** offer an abundance of health benefits. They're low in sugar and calories but high in fiber, manganese, and vitamins A, C, K, and B complex. They're also a good source of dietary fiber, which as you know by now, helps improve a dog's digestive system and fight obesity. Raspberries tend to keep your dog full for a longer period of time.

17 https://dogtime.com/dog-health/dog-food-dog-nutrition/59103-can-dogs-eat-peaches

Powerful antioxidants that can reduce the possibility of heart disease, cancer, diabetes, and arthritis, as well as minerals like potassium, manganese, copper, folic acid, iron, and magnesium, are also present in raspberries.

* **Spinach** contains vitamins A, B complex, C, E, and K. Vitamin A improves the skin, coat, and hair of the dog, in addition to preventing night blindness and stunted growth. B complex (B 2, 6, and 9) prevents hair loss and loss of appetite and improves reflexes and nerve control. Spinach also contains lots of plant proteins filled with amino acids, which help your dog thrive.
* **Strawberries** are a good source of vitamins C, B1, B6, and K, as well as fiber, potassium, omega-3 fatty acids, magnesium, iodine, and folic acid. These vitamins and minerals are helpful for immune system function and cell repair, and fiber aids in digestion, while fatty acids improve your pooch's skin and coat health. Not to mention, they're full of antioxidants. Unlike other foods, strawberries have the ability to whiten teeth.
* **Watermelon**, like most summer fruits on this list, is full of beneficial nutrients that are healthy for dogs, including potassium, vitamins A, B6, and C, and tons of fiber. Watermelon contains sugar, but the fiber content in the fruit insulates the sugar and prevents it from being released into the bloodstream too quickly. It's also a source of lycopene, an antioxidant that may help prevent cancer. Watermelon is low in calories, low in sodium, and free of both fat and cholesterol.

Chapter 5

Food Is Thy Medicine[18]

What do wild animals do when they get sick? Well, unlike Fluffy on the couch over there, animals in the wild don't have access to pharmacies or the wide array of treatments provided by their owners or veterinarians. So how do they survive without us?

Scientific evidence shows us that animals did (and still do) have knowledge of natural medicines. In fact, wolves have access to the world's largest pharmacy: nature itself. Zoologists and botanists are only just beginning to understand how wild animals use plants as medicine to prevent and cure illness.

Indigenous cultures have had knowledge of animal self-medication for centuries, and many folk remedies have come from noticing which plants animals eat when they are sick. Biologists and zoologists alike have witnessed animals eating foods not part of their usual diet, and realized the animals were self-medicating with natural remedies.

During a year-long observation period, a pregnant African elephant kept regular dietary habits throughout her long pregnancy but changed the routine abruptly toward the end of her term. Heavily pregnant, the elephant set off in search of a shrub that grew seventeen miles from her usual food source. The elephant chewed and ate the leaves and bark of the bush, then gave birth a few days later. The elephant, it seemed, had sought out this plant specifically to induce her labor. The same plant (a member of the borage family) also happens to be brewed by women in Africa to make a labor-inducing tea.

18 https://www.natural-wonder-pets.com/do-wild-animals-heal-themselves.html

Not only do many animals know which plant they require, they also know exactly which part of the plant they should use, and how they should ingest it. For instance, chimpanzees in Tanzania have been observed using plants in different ways. They eat shrubs called aspilia that produce bristly leaves, which the chimps carefully fold up then roll around their mouths before swallowing whole. The prickly leaves scrub parasitical worms from their intestinal lining.

Many animals eat minerals like clay or charcoal for their curative properties, which would explain why your dog eats dirt, mud, or even your clay pot. Colobus monkeys on the island of Zanzibar have been observed stealing and eating charcoal from human bonfires. Why? The charcoal counteracts toxic elements produced by the mango and almond leaves which make up their diet.

Some species of South American parrot and macaw are known to eat soil with a high kaolin content. Kaolin has been used for centuries in many cultures as a remedy for human gastrointestinal upset. A parrot's diet contains toxins because of the fruit seeds they eat. The kaolin clay absorbs the toxins and carries them out of the bird's digestive system, leaving it unharmed by the poisons.

So how do dogs know how to heal themselves? Some scientists believe that evolution has given them the innate ability to choose the correct herbal medicines. In terms of natural selection, wolves who could find medicinal substances in the wild were more likely to survive, while our generation of dogs seems to rely on what humans give them via their veterinarian's prescribed drugs.

Wild animals don't rely on synthetic drugs to cure their illnesses—the medicines they require are available in their natural environment. While animals in the wild instinctively know how to heal themselves, humans have all but forgotten this knowledge because of our lost connection with nature. Dogs, however, can still recognize and benefit from the wonders of nature, and in this chapter, you'll learn to use them.

In the last chapter, you got a quick course in what dogs can and can't eat. As you read through the list, I'm sure you noticed all of the vitamins and minerals that fruits and vegetables hold. In this chapter, we'll focus on using some of those ingredients in healthy healing recipes, in addition to introducing you to several other healing ingredients that work well with your dog.

* **Flax seeds**[19] are grown throughout the world but are mainly found in the United States and Canada. The plant produces tiny seeds that are tan or golden in color and have a nutty flavor. Flax seeds are safe and very nutritious for dogs. Whole seeds tend to sneak right through their digestive system, so ground flax seeds are a better option for recipes. It's best to keep these ground seeds very cold in the fridge or freezer to maintain freshness. Just a teaspoon a day can help make your dog's coat softer and shinier. Flax seeds also provide beneficial fiber, omega fatty acids, antioxidants, and alpha-linolenic acid. This latter compound is a plant-based form of omega-3 that offers an anti-inflammatory effect.
* **Chia** actually means "strength" in the Mayan language, which stands to reason, since these seeds were originally cultivated by the Mayans, Incas, and Aztecs. Chia seeds have been used to help alleviate skin irritation and joint pain. They're a plentiful source of protein, zinc, potassium, magnesium, phosphorus, iron, calcium, copper, and multiple B vitamins. They're also rich in antioxidants and alpha-linolenic acid. Plus, they balance electrolytes to help control stress and energy during prolonged play, training, or exercise. They're also great if you've got a dog that needs to lose a few pounds, as they swell up in the stomach and suppress the appetite.
* The Great Pumpkin, indeed. Just a single ounce of **pumpkin seeds** contains nearly nine grams of protein. They're also packed with folic acid, niacin, calcium, zinc, fiber, amino acids, copper, iron, phosphorus, manganese, magnesium, and potassium—not to mention vitamins A, E, and K. Looking for a rich source of inflammation-soothing omega-3? Check! Plus, your pups will also get a healthy dose of tryptophan to promote calming relaxation. Pumpkin seeds also contain an amino acid called cucurbitin, which can actually help control digestive tract parasites.

19 https://www.luckypuppymag.com/7-seeds-that-support-wellness-in-dogs/

Granola Bars with Chia, Flax, and Pumpkin Seeds

Granola bars are a great snack for dogs on a long hiking trip, or as a healthy treat to start the day. These granola bars are loaded with fiber, antioxidants, honey, and healthy fats, and are also great for the coat. They're delicious for humans, too!

Preparation Time: 20 minutes
Servings: 12–24
Equipment: sheet pan, mixing bowl, mixing spoon, parchment paper, measuring cup

Ingredients:

* 3¼ cups old-fashioned rolled oats
* ⅔ cups chickpea flour
* ⅔ cup honey
* ⅓ cup coconut oil, melted
* ⅓ cup chia seeds
* ⅓ cup pumpkin seeds
* ⅓ cup flax seeds

Directions:

1. Preheat the oven to 350°F and line a sheet pan with parchment paper.
2. In a large bowl, stir together oats, seeds, and flour.
3. Mix the honey and melted coconut oil in a measuring cup, then drizzle it over the oat mixture and stir until all the ingredients are completely combined.
4. Transfer the mixture into the prepared pan and press it down very firmly into an even layer.
5. Bake for 20–25 minutes, or until light golden brown on top.

6. Remove it from the oven and use the flat bottom of a small pan or dish to press down the granola again.

7. Cool it in the pan for 15 minutes, then carefully lift the granola out of the pan, and onto a cutting board. The best way to do this is to hold the parchment paper on one side and pull the pan from underneath on the other side with your cutting board under the granola bars.

8. Cut it into 12 large bars or 24 smaller bars. Place them onto a cooling rack to finish cooling completely.

9. Store them in an airtight container...if you don't eat them before the dog does!

* **Sunflower seeds** are brimming with nutrients, but they're also high in fat, so storing them in the refrigerator is the best way to discourage rancidity. They're a powerful source of phytosterols, which are thought to help reduce blood cholesterol levels, control inflammation, support heart function, and enhance immunity. They're also an excellent source of selenium, magnesium, vitamin E, manganese, phosphorus, copper, folate, and several B vitamins.
* **Quinoa** is sometimes referred to as a grain, but it's really a seed that's related to spinach and beets. It's also considered a complete protein, containing all nine essential amino acids. Quinoa is an excellent source of vitamins E and B6, folic acid, magnesium, manganese, zinc, copper, potassium, riboflavin, thiamin, and niacin. It's high in fiber and unsaturated fats, and it's extremely quick and easy to prepare, making it an awesome meal-topper. No matter how you decide to pronounce it, quinoa is a nutritional and protein-packed wonder seed.
* **Sesame seeds** have been cultivated since prehistoric times. They were originally brought to the United States from Africa around the seventeenth century. These seeds contain compounds called sesamin and sesamolin, which are beneficial fibers that help boost vitamin E reserves, control high blood pressure, and lower cholesterol. Sesamin has also been found to help shield the liver from oxidative damage. Sesame seeds are a great source of copper, and they give your dog a substantial dose of magnesium, manganese, selenium, calcium, phosphorous, and beneficial fiber. Sesame oil is also a great way to help heal rough, cracked paw pads.
* **Hemp seeds** contain omega-9 in the form of oleic acid; omega-6 in the form of linolenic and gamma linolenic acid; and the all-important omega-3 in the form of alpha-linolenic acid. They're full of amino acids, gluten-free protein, chlorophyll, and vitamins E and C. Hemp seeds help improve digestion, relieve joint pain and inflammation, and support cardiovascular wellness.

Since we ended our list with hemp seeds and I spoke of hemp oil, we can roll right into our next subject: oils that also heal. Just like you, your dog may not be getting all the necessary nutrition they need from eating their regular diet of processed foods. While standard dog food labels can say they come packed with plenty of essential nutrients, lots of them get lost in the process and from sitting on the shelf. These oils help with several issues and contribute to your dog's overall health and well-being.

* **Hemp oil** has a fatty acid profile closer to that of fish than any vegetable oil. Let's start by saying, yes, I know marijuana (cannabis) is on the "do not eat" list. While a dog eating your stash of weed can be frightening, when used properly, the marijuana plant does have several benefits for dogs. Hemp oil comes from the plant and is not psychoactive. It is completely free of THC and has none of the intoxicating aspects of cannabis. I'm sure you've seen the surge of new supplements on the market that are developed to treat a narrow range of health issues. However, they only scratch the surface of hemp oil's health benefits.
* **CBD oil** is great for dogs with anxiety of any kind, whether it's caused by thunderstorms, fireworks, a haircut, a trip to the veterinarian, and even separation anxiety. It also reduces inflammation. Studies show that dogs experience inflammation in the immune system as it fights real (or perceived) threats. Often inflammation exacerbates the underlying problem, as inflammatory cytokines are released for a variety of reasons. Especially in older dogs, this inflammation can become chronic, hurting joints and mobility. CBD oil has been shown to naturally realign inflammatory agents, so that your dog's immune system remains strong, but their joints are no longer swollen and painful. Along with reducing inflammation, CBD oil is a great alternative to pain pills. It targets the sensation of pain, which makes it great at treating all different kinds of ailments.

Hemp-Infused Coconut Oil

One of the most amazing medical benefits of hemp is its ability to help reduce and stop seizures. Studies show 5 percent of pets suffer from seizures. While most seizure treatments require a prescription from the vet, they are damaging to organs over time. Hemp oil is quite the opposite, with no side effects and 100 percent efficacy. The next recipe was created after multiple dog owners reached out to me and asked if I could make something their dogs' anxiety over Fourth of July fireworks., Fireworks are a big thing in Baltimore, but many dogs run away in fear of the noise they make. While there were several CBD products on the market at the time, none of them were working for these dogs. I made my own Hemp oil with the help of a holistic healer that worked perfectly. I don't know the laws on marijuana in your state, so I don't want anyone going to jail over dog treats! However, I will put the recipe here for those who feel comfortable making it. It really works, but please follow the recipe, as you want the balance to be right. Again, if you don't feel comfortable making the oil, find a good CBD oil to use.

Preparation Time: 25 minutes
Servings: 144
Equipment: heat-safe dish or sheet pan, small saucepan, sealable jar for storage

Ingredients:

* 12 oz. coconut oil
* 1½–2 g marijuana flower

Directions:

1. Preheat the oven to 225°F.
2. Place the marijuana flower in a covered heat-safe dish.
3. Bake (decarb) the marijuana flower for 20–25 minutes.
4. In a small saucepan, heat the coconut oil just until it liquefies and pour it into a container. Make sure it's not too hot!

5. Break apart the marijuana flower until it almost turns into a powder.
6. Add the powdered marijuana flower to the coconut oil and seal the container.[20]

When using this oil, the key element here is "less is more." Dogs are tiny in comparison to us, and the oil is powerful, so you don't need much. Having said that, 12 ounces yields 144 servings. You can put it directly on food, in recipes, or use an eye dropper to put it on the tongue.

The next recipe was created for a dog with severe anxiety. He had been to the veterinarian several times and had been prescribed several "human" anxiety meds, but nothing worked. It seemed as though the medications made him more anxious and confused. You could almost feel his anxious energy in the air, and he always appeared to be scared. I felt really bad for him and wanted to help, but he was always on the move, so I couldn't use an eyedropper. I knew he liked sweet stuff, so I made him these...and boy, did they work! They not only helped his anxiety in all situations but helped him become more relaxed altogether in his daily life. This one's for you, Louie!

---

20 One ounce of oil is twelve servings! Please pay attention to dose measurements when using in your creations!

CBD Brownies[21]

Dogs that experience anxiety display stress in different ways. Panting, shaking, aggression, destructive behavior and excessive barking are all common signs of anxiety. Pet owners may mistake such symptoms as their pet simply acting out due to boredom or other behavioral causes. However, if this occurs in common situations, like during a thunderstorm or when you leave the house, it can indicate that the dog is responding to anxiousness and stressful feelings. In Baltimore, one big issue was dogs running away during Fourth of July fireworks. We had an anxious dog, by the name of Louie. He was a rescue with an interesting history of anxiety-related escapes. When Fourth of July came around, it was his biggest nightmare, and ours, too. When he would get anxious, I'd feel so bad for him. It seemed like no medication would help—in fact, all of the medication made him worse. One year on the Fourth of July, we had run out of meds, and I knew it was going to be a big, loud firework show. I knew Louie liked sweet stuff, so I made him these awesome brownies. The change in him was amazing. He wasn't afraid and was actually happy. He even played with us during the fireworks!

Preparation Time: 30 minutes
Servings: 12
Equipment: mixing bowl, brownie pan, saucepan, coconut oil spray, mixing spoon

Ingredients:

* ½ cup coconut oil
* 1 cup honey
* 2 eggs
* ⅓ cup unsweetened carob powder
* ½ cup chickpea flour
* ¼ tsp. baking powder

21 https://www.petcarerx.com/article/how-to-know-if-your-dog-has-anxiety/

Directions:

1. Preheat oven to 350°F and grease an 8-inch square pan or brownie pan with coconut oil spray.
2. In a small saucepan, melt the coconut oil.
3. Remove it from the heat and allow it to cool, but not solidify.
4. In a medium-sized bowl, add honey, eggs, and measured CBD oil per serving. This means if you're using an 8-by-8-inch pan and you want 16 servings, you'd put 16 doses of CBD oil. If you are using a brownie pan with 12 cavities, put one dose per cavity. Make sense?
5. Stir in the carob powder, chickpea flour, and baking powder. Spread the batter into the prepared pan.
6. Bake it in the preheated oven for 25–30 minutes. Do not overcook.
7. Allow it to cool before removing it from the pan.
8. Enjoy!

Fish and krill oil both contain EPA and DHA, omega-3 fatty acids that help arthritis, improve memory, and have some anti-cancer effects. Fish-oil-based omega-3 fatty acids have a natural anti-inflammatory effect that can help reduce overall inflammation in the body. Krill oil comes from tiny shrimp-like organisms that rank a little lower on the food chain than fish. This makes krill oil less likely to be contaminated with mercury. Since dogs are omnivores that lean toward the carnivorous side, they absorb fish oils better than plant-based oils. Adding fish or krill oil to their daily regimen is definitely a great idea. Here's a way to make it fun...

Fish (or Krill) Oil Gummies

If you have a dog with arthritis, you know how uncomfortable it is for them. I had a client with severe arthritis, and nothing the vet prescribed seemed to work. I made these to help, and the dog loved them! The flavor of the gelatin made them hard to resist, and the effects were life changing. In combination with my meal plan and CBD treats, this dog was like a puppy again!

Preparation Time: 30 minutes
Servings: 12–24
Equipment: candy mold, coconut oil spray, small mixing bowl, mixing spoon, whisk, paper towel

Ingredients:

* 3¼ oz. unflavored gelatin
* ⅓ cup cold water
* ¼–½ tsp. of fish or krill oil
* Food coloring (sugar- and xylitol-free)

Directions:

1. Gather your ingredients, as well as a silicone gummy bear or worm mold. It can actually be any shape you want, as long as it's silicone for easy removal.

2. Spray a paper towel with coconut oil spray, then rub it lightly around the bear cavities in the mold to coat them with a thin layer of oil. This will prevent them from sticking to the mold.

3. In a small bowl, whisk the gelatin and the cold water. Let them sit at room temperature for 10 minutes to allow the gelatin to soak up the water and soften.

4. Place the bowl in the microwave and cook for 30 seconds, then whisk well.

5. Microwave for another 30 seconds and stir.

6. Add a few drops of food coloring and stir until your desired color is reached.
7. Pour the gelatin into the candy mold cavities.
8. Place the candy mold in the refrigerator for about 20 minutes to set the gelatin.
9. To remove the gummies, carefully push the sides away from the edges and toward the center, then pull them up and out of the molds.
10. Serve and enjoy (unused gummies can be frozen).

Extra-virgin **coconut oil** has become a popular choice for humans because it's a healthier alternative to more processed saturated and trans fats, and the same applies to dogs. Coconut oil has also been shown to help dogs lose weight, give them more energy, and offer relief to dry skin. Bonus: it will help improve your dog's bad breath, and it's also great for their coat! This cool candy recipe is not only delicious, but it's actually really healthy!

Coconut Booster Candy Bar

Preparation Time: 15 minutes
Servings: 4–8
Equipment: candy bar mold, saucepan, mixing bowl, mixing spoon, chef's knife

Ingredients:

* 1½ cup coconut oil
* 1½ cup unsweetened carob powder
* 1 orange (peeled, seeds removed)
* Candy bar mold that can be divided

Directions:

1. Melt the coconut oil over low heat until it liquifies, then turn off heat.
2. Chop the orange into small pieces. If possible, remove the membrane from the sides of the orange slices.
3. Empty the carob powder into a medium-sized bowl and stir in the coconut oil.
4. Fold in the orange pieces.
5. Pour the mixture into your molds and refrigerate until solid.
6. Enjoy!

* **Flaxseed oil** is high in alpha-linolenic omega-3s, which puts it in the same ballpark as wild fish when it comes to boosting heart health. Like many other healthy oils, flaxseed oil also helps with mobility for arthritic dogs, and can aid in blood pressure and kidney function.
* **Olive oil** is very popular and easily accessible at any grocery store. I lean toward extra virgin, as it has less olive taste, but either or is fine. Olive oil is wonderful for dogs' overall health

because it keeps their coats moisturized and shiny, improves their immune systems, and helps prevent and lessen the effects of cardiovascular disease and diabetes. It's great for pups whose systems aren't able to digest the omega-6 and omega-3 fatty acids well.

* **Sunflower oil** is a wonderful source of omega-6 fatty acids. It keeps a dog's skin moisturized, gives them energy, and keeps their heart healthy.
* **Safflower oil** is beneficial for dogs who lack unsaturated fats in their diets due to poor nutrition. If your dog suffers from seborrhea or keratinization disorders, they may benefit from linoleic acid present in safflower oil. The vitamin E present in safflower oil acts as an antioxidant and should be given to all dogs with the immune diseases and skin problems. Vitamin E also helps prevent cancer and improves oxygen circulation.

Herbs and Spices You and Your Dog Should Know About

Just like for us humans, herbs can add a healthy supplement to our dog's meals. Herbs and spices are most easily digestible for our canine friends when chopped very finely, or in the case of dried herbs, ground down to a powder. Definitely add these to their diet for optimal health.

* **Basil** is a lovely, leafy herb to add to your dog's raw diet. Basil has antiviral, antioxidant, and antimicrobial properties. Basil can help reduce the effects of arthritis, may help with inflammatory bowel disease and helps to repel insects.
* **Cinnamon** can help improve brain function, regulate blood sugar levels, lower blood pressure, and boost energy. It's also known for its antifungal and anti-inflammatory properties.
* **Coriander** is great for dental health and is a great bacteria growth inhibitor. It's antimicrobial and antifungal properties

can help in keeping infection at bay. As your dog eats it, it will also help fight any oral infections and improve their breath. It has various nutrients, including antioxidants to neutralize free radicals, vitamin A for healthy skin and vision, and vitamin K to help in blood clotting, among other roles. It can also help deal with digestion problems like gas, bloating, and upset stomach.

* **Dandelion** benefits the liver, skin, kidneys, and urinary tract, as well as the digestive and circulatory systems, and is rich in potassium. Dandelion is a strong but safe diuretic. Having said that, its beneficial to dogs with heart problems that cause an accumulation of fluids throughout the body, because it helps to excrete them. The diuretic property of dandelion also makes it an effective herb for dogs with urinary tract infections. It can help with an increase in urine flow by flushing out and eliminating the bacteria that are causing the infection. It's also a mild pain killer.
* **Garlic** has caused confusion over whether owners should use it for dogs, mainly because garlic is of the onion family and onion is definitely not good for dogs. But when used in small amounts, garlic is both safe and beneficial. Use only a fresh bulb, finely chopped—about a ¼ of a teaspoon once or twice a week for a forty-five pound dog is enough. Garlic is a natural antibiotic and antioxidant that helps boost immunity. It aids digestion and contains vitamins A, B complex, and C, proteins, and trace minerals. There is evidence that adding small amounts of garlic to a dog's diet can also assist with flea control.
* **Ginger** is an anti-inflammatory, antioxidant digestion aid that helps with blood circulation. It's also great for nausea or motion sickness, if you have a dog that doesn't like trips. If your dog has arthritis, ginger is great for easing the symptoms.
* **Mint** isn't just for fresh breath, it's also a good source of vitamins A and C. It also contains minerals such as calcium, copper, folate, iron, magnesium, manganese, niacin,

phosphorus, potassium, riboflavin, and zinc. It's calming elements can be used to soothe an upset stomach, reduce gas, and stave off nausea and motion sickness. Mint is a powerful antioxidant that's known for its antibacterial, antiviral, antimicrobial, and antifungal properties as well.

* **Milk thistle** strengthens and protects the liver to fight against toxins and is an extraordinary and extremely captivating herb in the treatment of dogs with liver diseases and associated disorders. Dogs with Cushing's disease have amounts of cortisol circulating in their blood. This stresses the liver, as it has to be strong enough to break down the extra amount of cortisol. Dogs suffering from Cushing's disease tend to have a high concentration of liver enzymes that, if left untreated, will cause liver disease. To strengthen the liver, dogs experiencing Cushing's disease can benefit enormously from supplementing their diets with milk thistle. In addition, various dogs suffering from inflammatory bowel disease tend to have pancreatic and liver irritation. So, if your dog suffers from IBD and additional liver/pancreatic problems, milk thistle helps with that, too.
* **Nettle** is rich in protein, vitamins A, C, K, B complex, and minerals such as calcium, phosphorus, iron, magnesium, and potassium. Due to its rich vitamin and mineral contents, nettle is a good herb to give to dogs who need an extra boost of trace minerals and vitamins in their diets. As an added benefit, nettle also has antibacterial, astringent, and antihistamine properties.
* **Oregano** helps with the inflammation associated with arthritis, is rich in antioxidants, and can be helpful when detoxing from a highly processed diet. It helps keep the coat and skin healthy. Oregano is high in flavonoids and is reported as an antimicrobial. This nontoxic herb has been used to help with digestive problems, diarrhea, and gas.
* **Parsley** is antibacterial, antioxidant, that helps to detox and is high in both fiber and protein. It contains a myriad of beneficial vitamins and minerals. Parsley helps fight bad

breath and aids digestion. It may help with UTIs and possibly with incontinence. You only need to add small amounts of parsley to your dog's meals, as very large daily amounts of parsley could become toxic to your dog.

* **Peppermint** is an antispasmodic—it also helps to stimulate circulation which can benefit your dog's skin and coat. Peppermint essential oil can treat motion sickness and arthritis and can even repel insects. It's also great for digestion.
* **Rosemary** is high in antioxidants, and great for heart function and the nervous system. Rosemary is rich in essential oils and contains flavonoids—a group of plant metabolites whose health properties consist of creating pathways that send signals to cells along with oxidative effects. Rosemary also contains phenolic acids, which are easily absorbed antioxidants in the gastrointestinal tract that prevent cell damage caused by free radicals. It has a calming, anti-depressant effect and helps fight anxiety and nervousness. It can be used to treat arthritis, stiffness and muscle pain, neuralgia, and sciatica, especially in older dogs.
* **Sage** is good for digestion and may help ease bloating and gas. It is used by herbalists to help support blood sugar regulation. It's good for cognitive function and helps reduce anxiety.
* **Thyme** is a good digestive aid that has antibacterial properties. It's good for skin and supports gastrointestinal health and brain function.
* **Turmeric** is a powerful antioxidant with anti-inflammatory properties. It contains a compound called curcumin, which is essentially its active ingredient. Curcumin is where turmeric's antioxidant, anti-inflammatory, antiviral, antibacterial, antifungal, wound-healing, and cancer-fighting properties come from. It can help fight diseases like arthritis, diabetes, cancer, liver disease, gastrointestinal issues, and more. Some call it "Cure-cumin" because of its long list of promising therapeutic and clinical uses.

Important Tip! When using medicinal herbs and spices in your dog's food, remember that they are smaller than us and require less amounts. For instance, when using powdered herbs and spices, use this rule:

* Toy breed: ⅛ teaspoon of powdered herb or spice
* Small dog: ¼ teaspoon of powdered herb or spice
* Medium dog: ½ teaspoon of powdered herb or spice
* Large dog: ¾ teaspoon of powdered herb or spice
* Extra-large dog: 1 teaspoon of powdered herb or spice

Chapter 6

Making Fresh Food

My whole premise for doing what I do is that I didn't like what I was seeing in commercial dog food. I questioned why we feed dogs the way we do, and what we could do to make it better. I also questioned why we as humans are trained to believe that dog food comes in a paper bag that we then sprinkle in a bowl for them. How did we get here?? Let's dive into the history of dog food for a minute.

So, the world's first commercial pet food was developed in 1860 by an electrician from Ohio named James Spratt. While traveling along the riverbanks of Northern London for work, he noticed dogs eating leftover biscuits that sailors ate on long trips and left behind. These inexpensive, imperishable crackers were made of flour, water, and salt. Spratt thought that dog owners were similarly in need of shelf-stable options for their pets, so he quit working as an electrician and got into the dog food business.

Spratt called his creation "Meat Fibrine Dog Cakes," and they contained a mix of wheat, vegetables, beetroot, and so called "dried unsalted gelatinous parts of Prairie Beef." He initially advertised his pricey food to the wealthy, and it was later represented as a full meal for dogs.

Spratt employed an aggressive advertising strategy, targeting health-conscious pet owners and dog show participants. In January 1889, he bought the full front cover of the first *American Kennel Club* journal to advertise his dog food. In promoting his product, he called upon a few old friends (read: rich, English country gentlemen) for testimonials touting the benefits of Spratt's dog cakes.

The American public was hooked and quickly traded the table scraps they'd been feeding their dogs for Spratt's biscuits. They also eventually latched onto his philosophy of "life stages" for dog food.

While business was booming for Spratt, he refused to disclose what meat was actually used in his products. In 1922, Ken-L Ration was introduced. It was government-approved as safe, with horse meat listed as a main ingredient. By 1941, canned food was so successful that producers were breeding horses just for dog food, slaughtering fifty thousand of them per year. However, when tin and meat were rationed during World War II and pet food was classified as nonessential, producers had to get creative. The combination of these imposed rations and pushback from animal lovers who were furious about the number of horses being killed every year for dog food created a golden opportunity to introduce a new wave in the pet food industry.

Several big names joined the race to find a new, more shelf-stable product, including General Mills, who acquired Spratt's US business in the 1950s, and The Ralston Purina Company, who began experimenting with the machines they were using for their Chex breakfast cereal.

In 1956, the first dry kibble was produced through a process called extrusion—a method used for manufacturing large quantities of shelf-stable foods. It works like this: wet and dry ingredients are mixed together to form a dough-like consistency, which is then fed into a machine called an expander. The dough is cooked under extreme pressurized steam and high temperatures before being extruded (or pushed) through a die cut machine and forming the small shapes we recognize as kibble today. The use of extrusion for commercial kibble production gained momentum throughout the 1960s and 1970s as companies used the technology to create new flavors and varieties.

In 1964, a lobbying group called "The Pet Food Institute" launched a series of ad campaigns to convince consumers that commercially prepared dog food was the only option to feed. The campaigns were hugely successful in convincing the American public that their dogs' diets should be kibble-based and were reminiscent of the early marketing strategies employed by James Spratt so many years before.

While this story is an awesome representation of an entrepreneur's American dream, it has become a nightmare for some dog parents, as more and more dog companies use lower-quality ingredients and unsafe practices to achieve a sellable product. Sounds familiar, right? Seems like humans have gone through a similar evolution, and the results of living on processed foods are equally as dangerous for both. As we've learned over the years, fast food isn't good for anything other than convenience.

When I started creating food for dogs, I had to ask myself "What did dogs eat *before* we got here?" Let's not forget, the first human bones from millions of years ago had her companion's bones right next to her. What did she feed *her* dog?

I would guess that they killed their food and hopefully shared it with each other. Well, killing random animals for food is no longer necessary, and humans don't eat much raw meat anymore. However, we do have access to fresh proteins, fruits, and vegetables at our grocery stores, and with the knowledge you acquire from this book, you, too, can become a dog chef!

First thing to acknowledge, is that you don't have to be a veterinarian or scientist to make dog food. To make food for anything that lives and breathes on the planet is to understand the importance of a balanced diet. As you learned about digestion and the immune system, you probably noticed similarities to your own digestive system and how it distributes minerals and nutrients. When food is processed, a lot of the essential nutrients are lost, resulting in little nutritional value. Feeding fresh food not only enhances the daily life of your dog, but it also prolongs it. All of our food is supposed to be "regulated" by the Food and Drug Administration (FDA). The FDA ensures the safety of that store-bought can of cat food, bag of dog food, or box of dog treats or snacks in your pantry.

The FDA's regulation of pet food is similar to that of other animal foods. The Federal Food, Drug, and Cosmetic Act (FFDCA) "requires that all animal foods, like human foods, be safe to eat, produced under sanitary conditions, contain no harmful substances, and be truthfully labeled. In addition, canned pet foods must be processed in conformance with the low acid canned food regulations to ensure the pet food is free of viable microorganisms."

There is no requirement that pet food products be pre-market approved by the FDA. However, the FDA ensures that the ingredients used in pet food are safe and have an appropriate function in the pet food. Many ingredients, such as meat, poultry, and grains, are considered safe and do not require pre-market approval. Other substances, such as sources of minerals, vitamins, or other nutrients, flavorings, preservatives, or processing aids, may be generally recognized as safe for an intended use.

These are the only existing guidelines for store-bought dog food and human food. So how do you know it's truly balanced?

There are several ways to make food for your dog. You've heard of the "BARF" diet, right? **BARF** stands for "Biologically Appropriate Raw Food" and "Bones and Raw Food." This way of feeding was brought to light by veterinarian and nutritionist Dr. Ian Billinghurst. The principle is to feed dogs the diet they evolved to eat: a raw diet composed of meats and greens that are fresh, uncooked, and wild. The genetic makeup of domesticated dogs supports this way of feeding—as stated before, dogs are essentially the same as their ancestors. The raw diet is high in protein, moderate in fat, and has minimal amounts of carbohydrates. This diet only includes muscle and organ meat, raw meaty bones, oils, fruits, and vegetables. I believe wholeheartedly in this theory and have seen it yield nothing but positive results. Before we get into the "how," let's talk about the "why."

Benefits of Feeding Your Dog Fresh Raw Food

There are numerous health benefits to feeding your dog a raw food diet. These include:

* Leaner, more muscular build
* Skin and coat health
* Cleaner teeth
* Less doggy odor
* Vibrant, calm energy

In addition, for all you hippies out there, feeding raw lowers your ecological footprint. A raw diet is more fully utilized by dogs' bodies, which equates to smaller stools.

Despite the many health benefits, one drawback associated with a raw diet is the risk of giving your dog an unbalanced diet. Many owners want to do it themselves, which is great. Unfortunately, most people don't know how to properly balance all of the micronutrients dogs need, which can lead to deficiencies and future health problems.

If your dog is new to raw food, the transition can often be accomplished within a week, but the key is to go slowly, as you would with any dietary change. Puppies can transition over the course of a few days, as they usually have healthier digestive systems than older dogs. The older the dog, the longer you should take to transition to the new raw diet. I recommend fasting your dog for at least a half-day before the first meal to ensure a good appetite. Feed them a little bit to see how they handle the fresh food. If all is well, continue replacing a little bit of their original diet with the new raw diet. If your pet experiences loose stools, wait until the stool is firm to continue the transition.

As a rule of thumb, I like to start crafting raw diets with a 30/70 split for smaller dogs and adjust accordingly. This rule allows you to make food in ten-pound increments in the beginning, and it starts with your food at 30 percent protein. Since organ meats are denser, you never want to use that as the main source of protein, but a little goes a long way. Organ meats help ignite a dog's primal drive and are very strong in smell, so they definitely make them excited about the food.

What's the other 70 percent? Glad you asked, as this is where the past two chapters come in. Each vegetable and fruit has a purpose in a dog's diet. While we will use them sparingly in the meals, the benefits will become evident almost immediately. This first meal is based on Greta's first raw meal but was made for a client switching her dog from kibble to raw food. The client was concerned her dog wouldn't like it, and her veterinarian was concerned about fresh food not having enough calcium. So after a talk with her veterinarian, I created this meal. Never use bone meal from the hardware store. Make it yourself! It can be done fairly easily and can be a money saver if you have some bones leftover from a prior meal. Poultry bones are recommended when making your own bone meal because they're easy to work with. These do splinter and break easily though, so you can also use beef bones (you may need to boil beef bones in order to make them pliable enough to cut). Just cut the bones into small pieces before they dry out, while they're still soft. Dry the small bone pieces for a few days, letting them harden and dry out thoroughly before placing them in a spice grinder and grinding them into powder. If you're looking for a general guideline to portion meals, the rule of thumb is to feed your adult dog 2 to 3 percent of their body weight when it comes to any food. For example, if your dog is fifty pounds, give them sixteen ounces or two cups of food per day. If your dog is twenty-five pounds, they'd need one cup, and so on.

What Are the Benefits of Feeding Raw Bones?

Raw bones offer many nutritional, physical, and mental benefits. They've been a necessity to dogs' well-being for thousands of years to help clean teeth, provide nutrients, and distract them from destroying things.

While many have opinions on this subject, I'm a huge fan of raw bones. They provide a slew of health and wellness benefits for dogs. They're completely digestible, perfectly natural, and an important part of a dog's daily diet.

You can grind bones and add them to your dog's food, or you can give them a large bone to chew on. If you're looking for something that will take a lot of time to chew, go with a hip or femur bone from large animals like cows and bison, which are filled with marrow.

Raw bones aren't just fun to chew, they also help clean your dog's teeth. Chewing raw bones is a great way to remove tartar off of dogs' teeth (especially the back molars) and keep them pearly white. Excess tartar buildup causes bad breath, cavities, and gingivitis. It can lead to expensive teeth scaling and extractions that can only be performed by a vet. When your dog chews on a raw bone, it helps stimulate saliva enzymes when given after a meal. All they need is ten to fifteen minutes to help remove trapped food particles from their teeth.

Raw bones are also a fantastic source of minerals like calcium phosphate, which aids your dog's proper growth. In fact, the calcium in raw bones is four times more digestible than store-bought supplements.

Generally, adding bone meal to your dog's food is a plus and is definitely recommended. If you don't have the equipment to grind bones, give a raw bone to your dog after they eat to help clean their teeth and obtain the other benefits of raw bones.

What Is a Balanced Meal for My Dog, and How Do I Create One?

In my many years of creating meals for dogs, I've learned that the key to a healthy dog is balance. Dogs are omnivores—they thrive on fresh protein and organ meat and benefit from fresh fruits and vegetables, just as we do. However, as you learned in chapter 3, we process food differently. It's important to understand these differences when creating food so you can adjust your levels accordingly.

Protein is a dog's main energy source, so when choosing a protein, you need to consider a few things... Humans eat seasonally. In the fall and wintertime, they eat hot proteins like beef or warming proteins like lamb, venison, or chicken. In the spring and summer months, they tend to do cold proteins like fish, rabbit, or duck (in addition to turkey, which is neutral and for all seasons). This philosophy applies to dogs as well—if you're feeding your dog a protein that's meant for a different season, it may explain a chronic skin issue your dog may be having.

Traditional Chinese Medicine states that an excess of yin (cold) can manifest as weakness, sluggishness, or even arthritis, while an excess of yang (hot) can manifest as irritability, dehydration, skin irritation, and digestive upset. A dog who is "hot" will often be warm to the touch and may pant even while resting. This canine may seem restless, constantly seeking out cool places. The eyes and/or skin may exhibit redness and irritation. I would characterize highly reactive or anxious canines, and those affected by chronic allergies, as "hot" dogs who should be fed a neutral or cold protein. A "cold" dog is one who may show signs of general fatigue, weakness, shortness of breath, exercise intolerance, and/or sluggishness. These dogs tend to seek out warm spaces. They may have urinary and/or fecal incontinence, joint pain that intensifies in cold weather, or stiffness that worsens with rest. If you lay your hand on a "cold" dog, you may notice that the back, limbs, and/or ears feel cold to the touch.

This also goes for the vegetables and fruits we use. Generally, there are seasonal vegetables for every country, but if you live in the United States, you can get anything you want all year round. When using vegetables and fruits in your dog's meals, understand their place in the diet.

I tend to go with the seasonal approach to give my dogs a variety. However, whatever fits for your dog is what you should do. Using this philosophy has helped me heal dogs with ailments for years.

When assembling your meals using this philosophy, you'll need to start with a baseline on how to portion your ingredients. I have developed a doggie food pyramid for you to follow based on how I go about making meal plans. There are many different methods, but this one is simple and easy to remember.

Cooling Foods:

* Apples
* Bananas
* Tomatoes
* Oranges
* Watermelon
* All leafy greens
* Broccoli and cauliflower
* Zucchini
* Mung beans
* Amaranth
* Wheat
* Seaweed
* Yogurt
* Peppermint

Neutral Foods:

* Rice
* Peas
* Lentils
* Large beans

Warming Foods:

* Ginger root
* Oats
* Quinoa
* Sesame
* Nuts
* Fennel
* Anise
* Carob
* Cumin
* All root vegetables
* Garlic

Doggie Food Pyramids

Large-Sized Dog

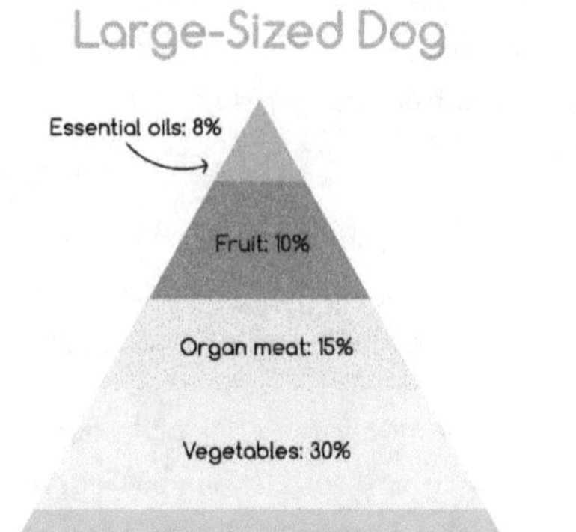

Medium-Sized Dog

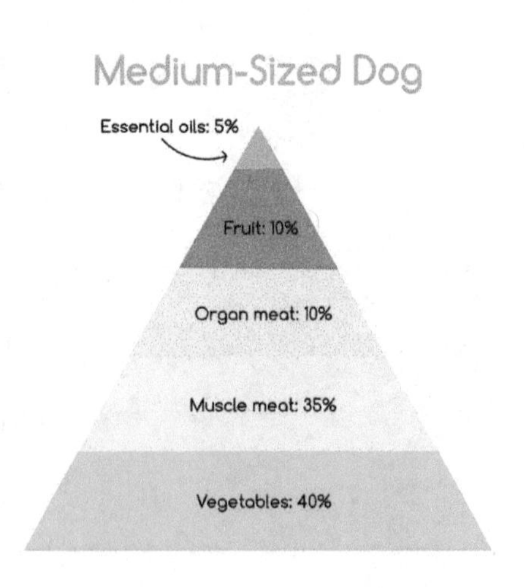

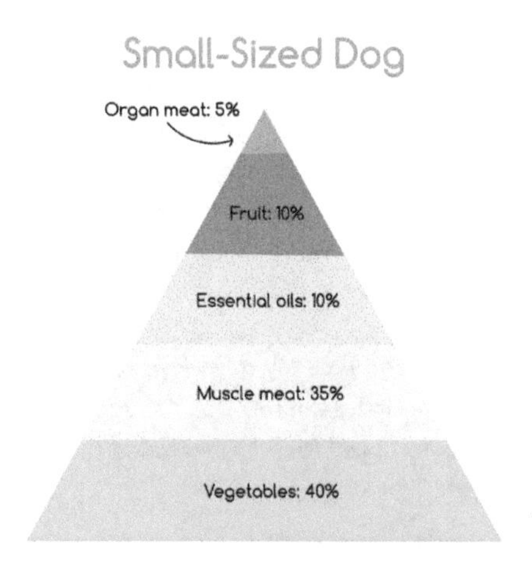

Preparing Your Veggies and Fruits

When creating fresh food, you have to remember a few things. A dog's digestive tract is shorter and more acidic than a human's, and they don't chew things to mush before swallowing. With this in mind, washing your vegetables and grinding them in the food processor allows dogs to digest them thoroughly and evenly in the digestive tract. There is really no need to cook them, as their stomach acidity is high enough to cook them. However, if you have an older dog with digestive issues, you can boil your veggies for a minute or so to aid the digestion process. Fruits need to be ground in the processor as well, but I don't recommend cooking them.

Preparing Your Proteins

When using raw proteins, it's imperative to use the freshest possible option. Clean the meat as you would for human consumption, and be careful if you're using raw chicken, due to salmonella concerns. While salmonella is rare to transmit to dogs, touching raw chicken and ingesting the bacteria can be dangerous. My dogs all love raw meat, but chicken can be hit or miss depending on the supplier. Because of this concern, I suggest lightly boiling your chicken meat. But don't cook the bones—cooked bones can splinter and cause damage to the esophagus or digestive tract, while raw bones break clean and digest well.

Fish can be served either way. My dogs *love* roasted salmon, but they like raw fish as well. I definitely don't suggest using cooked beef, as it's harder to digest (and I have found that it tends to cause diarrhea).

In addition, when you cook meat, you're removing essential nutrients the dog needs. Consider this when you're making meals, and if you choose to cook it, never overdo it. Lightly cook or boil to a rare. It's always good to make the food edible, not just pretty. Be sure to cut proteins into bite size pieces or grind them up completely.

Preparing Meals

Okay, so you know all about proteins, bones, veggies, fruit, herbs, spices, and oils. Let's move on to the fun stuff: making food! When I make food, I start with a basic recipe that embodies the food pyramid. I like to focus on vegetables by their purpose. I always go with one fruit, a leafy green, a cruciferous vegetable, a starch (potato, yams, or yucca), and a vegetable from the Apiaceae family (usually celery or parsnips, sometimes carrot, or a little of each). When picking your options, look through the lists in this book that include all useable fruits and vegetables and their benefits. Let your dog sniff the veggies and see what interests them. Don't be discouraged if they don't seem interested in the veggies or fruits just yet. When you mix them in the food, the smell of the protein will overpower them.

I want to conclude by touching on the importance of organ meat in your dog's meal. Organ meats ignite the primal drive your dog has. They can't resist food with organ meats...it's doggie kryptonite!!

Pick your vegetables (4), fruits (1), protein (1), and your organ meat (1), and head to the kitchen.

For the next step, you're going to need to figure out how much food you need to make. Let's think one week at a time for now. In order to do this, I use a simple math equation. Your dog's weight (or the weight you would like the dog to be if they need to trim down) multiplied by 0.03 will equal the amount of food your dog needs per day to maintain that weight. Let me give you an example:

If your dog is ninety pounds, you multiply their weight by 0.03, and their amount of food per day is 2.7 pounds (2 pounds, 7 ounces). If you have a thirty-five-pound dog and you multiply their body weight by 0.03, their amount of food per day is 1 pound and ½ an ounce. Follow the equations below to figure out your dog's ideal feeding weight.

________ x 0.03 = ________

Dog's Weight   Food per day in lbs.

Or, for a leaner dog, multiply their body weight by 0.02.

________ x 0.02 = ________

Dog's Weight   Food per day in lbs.

Now that you know how much your dog needs per day, use that percentage to figure out how much food you need to make balanced meals for the week. Since we used a thirty-five-pound dog in one of our examples, let's stick with that dog in this next example. If you remember, a thirty-five-pound, medium-sized dog eats up to 1 pound and ½ an ounce in a day. Seven days of that amount of food is 7 pounds and 3½ ounces. To make it easier, let's round the number up to 10 pounds to keep the measurements less confusing. According to the food pyramid, to make enough food for one week of balanced meals, this dog will need:

* **Fruit:** 10% = 1 lb. of fruit
* **Vegetables:** 40% = 4 lbs. of vegetables (1 lb. of each vegetable)
* **Essential oils:** 5% = ½ lb. (8 oz.) of oil
* **Organ meat:** 10% = 1 lb. of organ meat
* **Muscle meat:** 35% = 3½ lbs. of muscle meat

This will produce 10 pounds of food to be portioned into one-pound meals for the week. I understand that the meals were to measure to 1 pound and ½ an ounce. However, since we're using the heavier 0.03 component when making the equation versus 0.02, these meals can afford to be ½ an ounce lighter. If you want to be exact, make up the difference by adding more oil. In this case, you'd add 3 ounces of oil.

Does that make sense? Okay, let's try the same for that large dog that was ninety pounds. Since we're now talking about a larger dog, the food pyramid is a little different. With a larger dog, you need a bit more protein and fewer veggies. So if you remember, for a ninety-pound dog using our 0.03 formula, the daily food needed is 2 pounds and 7 ounces per day. That amount multiplied by 7 days equates to 17 pounds and 1 ounce of food for the week. As I did in the previous example, I will round the number to 20 to make the measurements easier, and still use four types of vegetables. According to the food pyramid for a larger ninety-pound dog, you will need:

* **Fruit:** 10% = 2 lbs. of fruit
* **Vegetables:** 30% = 6 lbs. of vegetables (1 lb. 8 oz. of each vegetable)
* **Essential oils:** 8% = 1 lb. (16 oz.) of oil
* **Organ meat:** 15% = 3 lbs. of organ meat
* **Muscle meat:** 45% = 8½ lbs. of muscle meat

Is it starting to make sense? Let's do it one more time for a small dog, and then you'll try with one of your dogs, okay? Alright, so let's say you have a ten-pound dog. According to our food pyramid, small dogs need less protein. Since organ meats are very dense, you need very little of them for a small dog. I usually don't use them unless necessary. If we use our 0.03 formula, the daily intake for this ten-pound dog is 3 ounces per day. I usually just round the number up to 4 ounces—the dog doesn't mind ☺

If you multiply that number by 7, it equals 28 ounces. Again, to make things easier, round it up to 32 ounces. With small dogs, you can usually add a bit more, but watch their weight and adjust as needed. So at 32 ounces (2 pounds) for the week, you can actually make enough food for a month and freeze the portioned meals. Since we're using that approach for the smaller dog, let's multiply the weekly number by 4 (weeks in a month) to see what we get. Using this calculation, this small dog will need 8 pounds of food per month. Here's the breakdown:

* **Fruit:** 10% = 8 oz. of fruit
* **Vegetables:** 40% = 2 lbs. (½ lb. of each vegetable)
* **Essential oils:** 10% = ½ lb. (8 oz.) of oil
* **Organ meat:** 5% = 4 oz. of organ meat
* **Muscle meat:** 35% = 2 lbs. 12 oz. of muscle meat

Okay, now it's your turn!

My dog ________________ weighs _________ pounds. If I multiply their weight by 0.03, I get ______________ as the amount of food. This number is the amount of food my dog needs per day in ounces and pounds. If I multiply that number by 7, I get the amount of food I need per week for fully balanced meals. By using the food pyramid provided for my dog's size, my balanced meals are as follows:

Fruit: ___% = ___ pounds ___ ounces of fruit

Vegetables: ___% = ___ pounds ___ ounces of vegetables (___ pounds ___ ounces of each veggie)

Essential oils: ___% = ___ ounces of essential oils

Organ meat: ___% = ___ pounds ___ ounces of organ meat

Muscle meat: ___% = ___ pounds ___ ounces of muscle meat

With this formula done, you have assembled your plan of action. The next step is taking the vegetables, fruits, and meat and planning out your first recipe. I want you to write down the ingredients in the spaces provided on the next page and follow the instructions. I've already figured this out for you, because the next steps never change as part of your routine. Get familiar with it, you'll be doing it a lot.

One more thing: remember to occasionally give your dog a raw bone after meals. Grinding bones in your food processor may break it, so your best bet would be to give them fresh raw bones after a meal.

My First Recipe

Preparation Time: 45 minutes
Servings: 7
Equipment: food processor, chef's knife, large mixing bowl, vacuum sealer, vacuum sealer bags

Ingredients:

* ________lb(s). protein ______________________________
* ________lb(s). organ meat __________________________
* ________lb(s). fruit _________________________________
* ________lb(s). vegetable ___________________________
* ________lb(s). vegetable ___________________________
* ________lb(s). vegetable ___________________________
* ________lb(s). vegetable ___________________________

Directions:

1. Wash, chop, and grind all vegetables in the food processor and put them in a large bowl.
2. Remove seeds from fruits as necessary, then chop and pulse them in food processor.
3. Put the fruits together with the vegetables in the bowl and mix.
4. Add essential oil to the bowl and toss the fruit and vegetables until the oil is evenly dispersed.
5. Finely chop your proteins and organ meats as necessary into bite-size pieces, unless you're using a ground protein.
6. Add your proteins and organ meats and mix everything until all ingredients are thoroughly combined.

7. Portion your meals by vacuum sealing individual bags for each day and freeze them.

8. You did it! Congratulations!! Woo hoo!!!

Watch your dog closely as you adjust the recipes to fit your needs. Pay attention to the vitamins and nutrients that the ingredients provide to your dog. Be aware of any symptoms of vitamin deficiencies or changes in your dog's behavior, appearance, and weight, and adjust your meals accordingly. This takes time to get right, but with confidence, you can get it in no time! I've provided a few sample meals for you to try working with. Feel free to tweak the meals to fit your dog's needs.

Greta's Favorite Raw Chicken Dinner

This first meal is based on Greta's first raw meal, but it was also made for a client switching her dog from kibble to raw food. The client was concerned her dog wouldn't like it, and her veterinarian was concerned about fresh food not having enough calcium. So after a talk with her veterinarian, I created this meal.

When I made Greta's first meal, I used a whole raw chicken and filled the carcass with all of the ingredients. I didn't cut it up or anything, I served it just like that in the bowl. While I enjoyed watching her eat it, it was a bit too much. It was messy, —she dragged the dead chicken all over the couch! I had to find a safer, more efficient way for her to eat these creations I was making for her. I decided to try ground chicken, which allowed me to portion the meals evenly and be more balanced. Sometimes I'd use chicken legs and chop them into pieces so she could still have raw bones. After switching to this method, I really saw the benefits of raw food. She really started to look and behave like a different dog. She was more alert and focused, had an amazing coat and shiny white teeth, and was more active for longer periods of time. I would get on my bike and grab her leash to take her to Central Park and we'd loop the park twice with her running next to me the entire time...and she would still want a walk when we got home!!

Preparation Time: 30 minutes
Servings: 7
Equipment: food processor, chef's knife, large mixing bowl, mixing spoon, vacuum sealer, vacuum sealer bags

Ingredients:

* 3 lbs. raw ground chicken
* 1 lb. broccoli
* 1 lb. kale
* 1 lb. apples

* 1½ lb. celery
* 1 lb. baby carrots
* 1 can chickpeas, rinsed and drained
* 1 lb. sweet potato
* 1 cup olive oil
* 1 cup Greek yogurt

Directions:

1. Wash, chop, and grind all vegetables in the food processor and put them in a large bowl.
2. Remove seeds from the apples, then chop and pulse them in the food processor.
3. Put the apple pieces in the bowl with the vegetables and mix.
4. Add chickpeas and olive oil to the food processor and grind until the mixture reaches the consistency of hummus.
5. Add the chickpea mixture, ground chicken, and yogurt to the bowl and mix until all ingredients are thoroughly combined.
6. Portion your meals and freeze.

Buddy's Beef Stew

I met buddy's mom at a doggie function I was catering. She was a news reporter doing a story on the event, and she interviewed me about dog food. Her dog didn't have any issues at the moment other than being a bit itchy. "Buddy eats everything!" she said and proceeded to tell me that he was her first dog as an adult. She loved Buddy like he was her child and wanted the best for him. "I want Buddy to live forever," she said. I suggested raw food for Buddy, but she was unsure of its safety, as it was a new concept. A week after getting the food, she wrote me an email telling me how much Buddy liked it. With dry food, Buddy would visit the food bowl throughout the day to finish his meals. He liked the raw food so much that he licked the bowl *clean* in less than five minutes!

Preparation Time: 30 minutes
Servings: 7
Equipment: food processor, chef's knife, large mixing bowl, mixing spoon, vacuum sealer, vacuum sealer bags

Ingredients:

* 2 lbs. 80/20 lean ground beef
* 1 lb. beef liver
* 6 tbsp. bone meal
* 1 lb. broccoli
* 1 lb. kale
* 1 lb. apples (green or red)
* 1½ lb. celery
* 1 lb. sweet potato
* 1 cup olive oil

Directions:

1. Wash, chop, and grind all vegetables in food processor, then put them in a large bowl.
2. Remove seeds from apples, chop, and pulse them in food processor, and put apple pieces in the bowl and mix.
3. Put the liver in the food processor, grinding it into a paste while adding the bone meal.
4. Add the liver mixture to the bowl.
5. Add the ground beef to the bowl and mix until all ingredients are thoroughly combined.
6. Portion your meals and freeze.

Indu's Chicken

I got a request to meet a client who wanted to switch to fresh food. Her concerns were based on the same fears that inspired me to write this book. There were a lot of dog food recalls happening, and she didn't feel safe using store-bought food anymore. Upon meeting the dog, I noticed two things: her beautiful eyes and how uncomfortably overweight she was. Indu was a younger dog with joint issues, so I knew the joint pain was mainly attributed to her weight, not her age. If she continued on this path, there's no doubt she would have caused further damage. She was so overweight, she seemed to break into a pant from just standing. When creating this meal, there were several factors I had to take into consideration. The dog liked chicken, but the owner wasn't comfortable feeding raw, so this was catered to both of their needs. Her family was from India, and this was their first dog. Staying true to the family's religion, I chose not to use beef, as cows are sacred in India. I also wanted to use vegetables that were native to their country, as part of Indu's problem was that her owner's parents kept giving her table scraps (which she loved ☺). So, I crafted the meal as a dish that Indu would feel like the family would eat. Because of the beautiful colors and ingredients, it looked like an authentic Indian dish. Ten years later, she's thriving on this same recipe!

Preparation Time: 30 minutes
Servings: 7
Equipment: food processor, chef's knife, large mixing bowl, mixing spoon, vacuum sealer, vacuum sealer bags

Ingredients:

* 5 lbs. chicken thighs
* 1 lb. chicken liver
* 6 tbsp. chicken bone meal
* 2 lbs. cauliflower
* 2 lbs. green cabbage
* 1 lb. celery
* 1 lb. yucca root
* 1 lb. frozen peas and carrots
* 1 lb. pears
* 1 cup olive oil
* 1 can chickpeas, drained and rinsed
* ⅛ cup turmeric

Directions:

1. Clean the chicken thighs and livers and place them in a medium to large pot. Fill it with water and boil for 20 minutes, then drain and remove it from the heat.
2. Wash, chop, and grind all of the fresh vegetables (except the peas and carrots) in the food processor and put them in a large bowl.
3. Remove the seeds from the pears, then chop and pulse them in the food processor.
4. Put the pear pieces in the bowl with the vegetables and mix.
5. Put the chicken thighs and liver in the food processor and grind them into a paste while adding the bone meal.
6. Add the mixture to the bowl.
7. Add the frozen peas and carrots to the bowl and mix until all of the ingredients are thoroughly combined.
8. Add the chickpeas, olive oil, and turmeric to the food processor and grind until it reaches the consistency of hummus.
9. Add the chickpea mixture to the bowl and mix until the meal has uniform yellow color.
10. Portion your meals and freeze.

Coco's Salmon

This next recipe was created after meeting a client with a little dog named Coco. She was a Maltese that was supposed to be white, but she appeared to be brown when I met her and had all of this brown goop coming out of her eyes. Her owner also told me that Coco always had diarrhea. Turns out, that goop was toxins from the "vet-prescribed" food that was given to her for her liver issues and hair loss. She was on tons of medications and seemed lethargic. Even with the vet-prescribed food, nothing seemed to change her condition. When I asked her owner what she wanted the food to do for her dog, she said, "Well, we just want her healthy and want her to be white again." I looked at the dog and asked, "What does she like?" They mentioned that she loved fish, so I created this meal plan for her. Within a week, she was like a new dog. She no longer had that "doggy odor," and the eye goop stopped. After a month, she was a little white dog with tons of energy!

Preparation Time: 30 minutes
Servings: 7
Equipment: food processor, chef's knife, large mixing bowl, mixing spoon, sheet pan, parchment paper, medium saucepan, vacuum sealer, vacuum sealer bags

Ingredients:

* 5 lbs. whole salmon fish, cleaned (bones and scales removed)
* 2 lbs. cauliflower
* 2 lbs. purple cabbage
* 1 lb. celery
* 1 lb. yucca root
* 1 cup olive oil
* 1 lb. green apples
* 2 cups cooked quinoa

Directions

1. Preheat the oven to 350°F.
2. Place the whole salmon on a sheet pan lined with parchment paper and bake for 20 minutes. Then remove it from the heat and allow it to cool.
3. In a medium saucepan, add quinoa and bring 4 cups of water to a boil for 15–20 minutes. Then remove it from the heat.
4. Wash, chop, and grind the cauliflower, cabbage, and celery in the food processor and put them in a large bowl.
5. Remove the seeds from the green apples, and chop and pulse them in the food processor. Then put the apple pieces in the bowl with the vegetables and mix.
6. Break apart the salmon in the food processor with skin and bones. Grind it to a paste.
7. Add the salmon mixture and olive oil to the bowl.
8. Drain the quinoa and add it to the bowl as well.
9. Mix everything thoroughly until the quinoa is equally distributed.
10. Portion your meals and freeze.

Cooper's Turkey

When I got the call for Cooper, his owner mentioned that he was allergic to everything. I didn't believe her, so I asked her to send me the allergy test from her veterinarian. When I saw it, I realized that this was going to be the most difficult dog I'd ever made food for. His list of allergens was three pages long, and every food they tried with him didn't work. The only protein that didn't cause a negative reaction was turkey, so that was my starting point. The only five ingredients on his "can have" list were oddly in the group of vegetables to make a balanced meal. Combined with the only oil he wasn't allergic to, this made an awesome balanced meal for him. To this day, Cooper is still thriving, with no reports of *any* allergic reactions! While this meal is best raw, it can be cooked as well.

Preparation Time: 30 minutes
Servings: 7
Equipment: food processor, chef's knife, large mixing bowl, large cooking pot, vacuum sealer, vacuum sealer bags

Ingredients:

* 6 lbs. ground turkey
* 2 lbs. strawberries
* 2 lbs. green cabbage
* 2 lbs. yucca
* 2 lbs. cauliflower
* 2 lbs. celery
* 1 cup safflower oil

Directions:

1. If making the raw version of this recipe, skip steps 2, 3, and 4.

2. In a large pot, add the ground turkey and fill it with water until it covers the turkey.

3. Bring it to a boil, then set it to low heat for 25 minutes.

4. Turn off the heat and allow it to cool for a bit, then drain the water and set it aside.
5. Wash, chop, and grind all vegetables in the food processor, then put them in a large bowl.
6. Wash the strawberries and pulse them in the food processor, then add them to the bowl.
7. Add the ground turkey to the bowl.
8. While hand mixing the ingredients, add the safflower oil. Mix thoroughly.
9. Portion your meals and freeze.

Sushi Rolls

Making sushi at home is surprisingly simple and a great way to introduce your dog to raw proteins. We'll start with sushi for beginners, or sushi rolls. With just a little practice, you can make sushi rolls at home that are as dazzling to look at as they are delicious for your dog to eat. These are also shareable!

Preparation Time: 25 minutes
Servings: 2–5
Equipment: bamboo mat, plastic wrap, chef's knife, saucepan, cutting board

Ingredients:

* 1–1.5 sheets of nori (seaweed paper)
* 1 cup sushi rice
* 1½ tsp. honey
* Raw steak or salmon
* Celery and carrots

Directions:

1. In a small saucepan, add the sushi rice and 2 cups of water over high heat.
2. Bring it to a boil, uncovered. Once it begins to boil, reduce the heat to the lowest setting and cover. Cook it this way for 15 minutes.
3. Drizzle the honey over the rice, fluffing it with a fork.
4. Remove it from the heat and let it stand, covered, for 10 minutes.
5. Place the bamboo mat on a cutting board so the bamboo strips are running horizontally to you. Spread a strip of plastic wrap over the bamboo mat, then place a sheet (or a half sheet) of nori on the plastic wrap.

6. Spread a thin layer of sushi rice over the nori, but don't use too much. With a little practice, you'll get a feel for how much rice to use. Spread about 1 cup of rice per 1 whole sheet of nori, leaving a small space at the top edge of the nori to seal up the roll.

PRO TIP: Wet your fingers as you spread the sticky sushi rice over the nori.

7. Add your ingredients toward the center of the rice-covered nori. Don't overpack it. If you're making a roll with the rice on the outside, simply turn the nori over and place it rice-side down on the rolling mat, then add your ingredients to the top of the nori.
8. Gently lift the bottom of the mat up and over the sushi, then press and shape the ingredients into a tube.

9. Roll with pressure so you get a firm roll, and until just an inch of nori shows at the top. Then seal the edge of the nori with a little cold water.

10. Firm it up by squeezing the mat around the roll until it feels uniformly snug. Be careful not to squeeze so hard that the ingredients are smashed or come oozing out the sides. It's a fine line, but with practice, you'll get the feel of it.

11. Slice the sushi roll in half with a sharp knife on a cutting board. Then cut each half into thirds, leaving you with 6 pieces of sushi. To get a nice clean cut, wet the knife with water each time you make a slice.

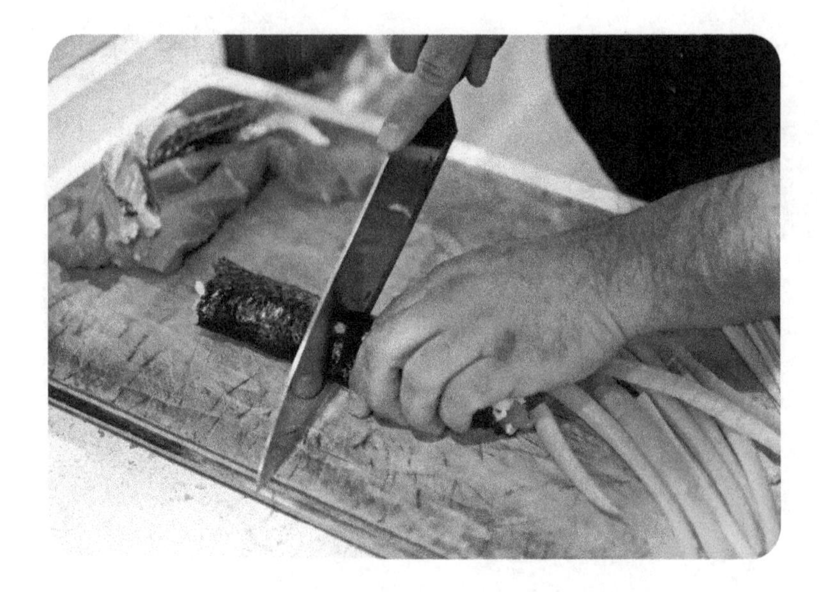

12. Line the slices of sushi up on platters or sushi plates, or toss 'em like treats.
13. Enjoy!

While these recipes are a basic start, they embody the balanced thought process that can be used as a building block for your success in making your very own recipes. Watch your dog closely as you adjust the recipes to fit their needs. Pay attention to the vitamins and nutrients that the ingredients provide your dog. Be aware of any symptoms of vitamin deficiency, or changes in your dog's behavior and weight, and if for any reason you feel your dog seems ill, contact your veterinarian immediately. Check your local pet store for any available supplements for natural diets that will offer a baseline of the daily vitamins your dog needs and incorporate them into your regimen.

Chapter 7

Treats, Sweets, and Holiday Eats

Holiday Thumb Cookies

The best thing about the holiday season for me is the cookies! When I was a kid, we'd get tons of cookies at Christmastime, but my favorite ones were thumb cookies. I used to love when my Mom would put them in my lunch. The filling would harden up in the cold and taste like candy. My childhood dog Heidi loved them, too. Since the ones for humans are filled with sugar and not good for dogs, I made this rendition of my favorite holiday cookie at my bakery in a Christmas cookie sampler. They're super healthy and sweetened with honey to add to the deliciousness. They're a must for your next doggie Christmas party!

Preparation Time: 45 minutes
Servings: 36
Equipment: mixing bowl, mixing spoon, saucepan, sheet pan

Ingredients:

* 3 cups chickpea flour
* 1 cup water
* ¾ cup honey
* 1 egg
* 8 oz. fresh cranberries

Directions:

1. Preheat your oven to 350°F.
2. In a medium-sized bowl, thoroughly mix the water, egg, and ¼ cup of honey.
3. Slowly mix in the chickpea flour until it becomes a dough. If the dough is too sticky, add a bit more flour.
4. Roll the dough into 1-inch balls and put them on your sheet pan, then set them aside.
5. In a small saucepan, add cranberries. Then add water until it covers the cranberries.

6. Turn the heat up to medium and let the cranberries boil down until they become a paste.
7. Add the honey to the saucepan and stir.
8. Remove the mixture from the heat and let it cool. While that cools, take your thumb and press down in the center of each of your dough balls.
9. Fill those indents with the cranberry reduction.
10. Bake them for 15 minutes, then allow them to cool before feeding them to your dog.
11. Enjoy!

Liver Pill Pâté

If you have a dog that takes medication, you may understand how difficult it is to get them to take a pill. When I was a kid, we'd try everything to get our dog to take pills. But every time, she would just eat what the pill was wrapped in and spit out the pill itself. This particular recipe was designed specifically for this purpose. By using liver, you're igniting the dog's primal drive. That smell overpowers any other smell for them, and the pomegranate adds an extra flavor and antioxidant component.

Preparation Time: 20 minutes
Servings: 15
Equipment: saucepan, small skillet pan, food processor

Ingredients:

* 3 cups water
* 1 lb. beef or chicken liver
* ¾ cup butter, softened
* ¼ tsp. salt
* 1 rosemary sprig

Directions:

1. In a medium saucepan, combine water, liver, and pomegranate juice, and bring it to a boil.
2. Reduce the heat to low, then add the rosemary leaves and cover the mix. Let it simmer for about 20 minutes, or until the liver is cooked and tender.
3. Remove it from the heat and drain it. Be sure to discard of the onions and any hard portions of the liver.
4. Place the cooked liver in a blender or food processor, and process it until smooth.
5. Chill for 1 hour before serving.
6. Take the pill, cover it with pâté and serve.

Glazed Pumpkin Oat Cookies

These oatmeal cookies are based on one of my old-school favorites. I used to *love* glazed oatmeal cookies with raisins. As you know, raisins are a no-no for dogs, so I decided to change the recipe up a bit. By adding pumpkin, this cookie becomes a powerhouse of fiber and flavor, and dogs love the glaze! These are great if you want to give them a treat at the dog park!

Preparation Time: 45 minutes
Servings: 36
Equipment: sheet pan, rolling pin, 2 mixing bowls, whisk, mixing spoon, cookie scoop or spoon, parchment paper

Ingredients:

* 3 cups chickpea or barley flour
* 1 cup old-fashioned rolled oats[22]
* ½ cup canned pumpkin puree
* ¼ cup water
* ½ tsp. baking powder
* 1 tsp. cinnamon

For the glaze:

* ½ cup tapioca starch
* 2 tbsp. honey
* ¼ cup coconut milk

Directions:

1. Preheat the oven to 350°F.

2. In a large bowl, mix the water and the pumpkin puree.

3. Add the flour, oats, baking powder, and cinnamon, and mix until the ingredients are combined.

22 Note: Steel-cut oats will not work in this recipe. Use flat, old-fashioned rolled oats!

4. Scoop the batter using a cookie scoop or a tablespoon and place the scoops onto a baking sheet lined with parchment paper.
5. Bake them for 15 minutes, or until the edges are golden and the cookies are set. They will be soft!

For the Glaze:

1. While the cookies are baking, in a small bowl, whisk together the tapioca starch and honey until the mixture is smooth.
2. Slowly add the coconut milk per tablespoon until the consistency is met.
3. Drizzle the glaze over the cookies in whatever pattern you like, then allow them to dry.

Birthday Cupcakes

Yep, dog birthdays are a "thing," and just like we humans, dogs love cupcakes!! While regular cupcakes could do some damage to your pup, these are not only delicious, but they're also super healthy! With the antioxidant powers of honey and the digestive boost from the yogurt icing, these are a must for any puppy pawty!

Preparation Time: 60 minutes
Servings: 12
Equipment: 2 mixing bowls, fork, food processor, muffin pan, piping bag, decorating tip, cupcake liners

Ingredients:

* 1½ cups flour
* ½ lb. sweet potato
* ½ cup canola oil
* ½ cup honey
* 2 small eggs
* 1 heaping tsp. baking powder

For the Icing:

* 16 oz. full-fat Greek yogurt (low-fat yogurt doesn't work!)
* 8 oz. tapioca starch

For the Sprinkles:

* 1 cup shredded coconut
* Food coloring
* 4 Ziploc bags

Directions:

1. Preheat the oven to 350°F.
2. In a large bowl, combine the eggs, honey, and oil. Mix well.
3. Cut the sweet potato into small pieces, then grind them in the food processor into fine pieces.

4. Add the sweet potato pieces to the bowl and stir.
5. Add the flour and baking powder to the bowl and stir well with a fork. Be sure to break up any lumps!
6. Put liners of your choice in your cupcake pan.
7. Fill cupcake liners ¾ of the way full. Be sure not to put too much or they will spill over.
8. Bake your cupcakes for 20–25 minutes.
9. Remove them from the heat and allow them to cool.
10. In a medium-sized bowl, combine the yogurt and tapioca starch. Mix them well, until the consistency is like cake frosting.
11. Fill a piping bag with the frosting, then refrigerate it for about 10 minutes.
12. Fill each Ziploc bag with ¼ cup coconut flakes and add 8–10 drops of the primary colors in each bag.
13. Zip the bags and shake them vigorously until the colors are evenly distributed, then open the bags and allow them to dry for 5 minutes.
14. Use the widest open bit of your bag and hold it a ½-inch away from the cupcake. Apply pressure to the bag and squeeze out the icing until it creates a circle that expands to the edge of the cupcakes, then release pressure and pull upwards. The tops of the cupcakes should look like a Hershey's kiss.
15. Mix the colored sprinkles together in a separate Ziploc and sprinkle them on top of your cupcakes.
16. Happy birthday!!!

Carob Chip Cookies

Who doesn't love chocolate chip cookies? Dogs!! Due to their chocolate intolerance, I designed this recipe with dogs in mind. These are a twist on the classic but made with an interesting alternative to chocolate: carob. In my opinion, it tastes even better! These are great for any occasion, as well as shareable with humans. Be careful—they're delicious, so make sure you leave some for the dog!

Preparation Time: 35 minutes
Servings: 24–26
Equipment: sheet pan, mixing bowl, mixing spoon (wooden preferred), parchment paper or coconut oil

Ingredients:

* 12 oz. unsweetened carob chips
* 1 egg
* ½ cup honey
* ½ cup water
* 1 tsp. baking powder
* 4 cups chickpea or barley flour

Directions:

1. Preheat the oven to 325°F.
2. In a large bowl, whisk together the egg, honey, and water.
3. With a wooden spoon, stir in the flour and baking powder. Mix well until it's a cookie dough. Add more water or flour as needed, as the dough should not be sticky.
4. Form 1-inch balls and place them 2 inches apart on a parchment-paper-lined sheet pan. If you don't have parchment paper, you can grease the cookie sheet with coconut oil.
5. Press the balls flat, about ⅛ to ¼ of an inch thick.

6. Bake the cookies for 15–20 minutes. For crispier cookies, bake them at 350°F.

7. Allow them to cool, then enjoy!

I urge you to take a bite of these delicious treats. If you're allergic to chocolate, they're a great, healthy alternative!

Apple Heart Cookies

In my house, Valentine's day is a worthy celebration of the love we have for our dogs, and we spoil them accordingly. I like to make these apple heart cookies, as they remind me of the conversation heart candies. Unlike the candy version, these are great for a dog's digestive system and coat!

Preparation Time: 30 minutes
Servings: 42
Equipment: sheet pan, 2 mixing bowls, rolling pin, mixing spoon (wooden preferred), heart-shaped cookie cutter

Ingredients:

* 1 cup fresh apple puree, or organic apple sauce (no sugar, just apples!!)
* 1 cup flaxseed meal
* 3 cups chickpea flour
* 1 tbsp. cinnamon

For the Frosting:

* ½ cup tapioca starch
* 2 tbsp. honey
* ¼ cup coconut milk
* Red food coloring

Directions:

1. Preheat the oven to 350°F.
2. In a large bowl, combine the flour, cinnamon, and flaxseed meal.
3. Add the apple puree to the bowl, then stir the mixture with a wooden spoon until it forms a dough.
4. On a clean, lightly floured surface, roll the dough into a ¼-inch thick sheet.

5. Use a heart-shaped cookie cutter to cut the shapes into the dough and move them onto the parchment-paper-lined sheet pan.

6. Bake the cookies for 15–20 minutes, until they're golden brown.

7. Remove the cookies from the heat and allow them to cool.

For the Frosting:

1. In a small bowl, add the honey, milk, and food coloring, then stir in the tapioca starch. Add more milk if needed to reach the consistency of pancake syrup.

2. Drizzle the glaze over cookies or dip the cookies in the glaze.

3. Allow the cookies to dry for about 1 hour, until the glaze hardens.

4. Enjoy!

Deviled Eggs

Deviled eggs make a great appetizer or hors d'oeuvre for any situation. They say eggs are a perfect protein, and I agree! Great for brain, bone, and eye health, as well as a shiny coat. Deviled eggs are a treat that can't be beat!

Preparation Time: 15 minutes
Servings: 3–6
Equipment: mixing bowl, saucepan, fork, teaspoon, serving platter, paper towels

Ingredients:

* 3 eggs
* ¼ cup plain yogurt
* 2 tbsp. fish or krill oil
* Chia seeds

Directions:

1. Place the eggs in a single layer in a saucepan and cover them with water (enough that there's 1½ inches of water above them).

2. Heat them on high until the water begins to boil, then cover them, turn the heat to low, and cook them for 1 minute.

3. Remove them from the heat and leave them covered for 14 minutes, then rinse them under cold water continuously for 1 minute.

4. Crack the eggshells and carefully peel them under cool running water, then gently dry them with paper towels.

5. Slice the eggs in half lengthwise, placing the yolks into a medium-sized bowl, and placing the whites on a serving platter.

6. Mash the yolks into a fine crumble using a fork, then add the yogurt and fish oil. Mix well.

7. Evenly disperse heaping teaspoons of the yolk mixture into the egg whites, then sprinkle them with chia seeds and serve.

Honey Roasted Chickpeas

Chickpeas are used worldwide by many cultures. As you've seen in this book, I use mostly chickpea flower. That's because it's gluten free and provides a good serving of protein and fiber. In addition to its many health benefits, it's very versatile and delicious. I created these treats for dog training, as they can easily fit in your pocket. Dogs love them—they're a great shareable snack that's hard to resist. The mix of savory and sweet make these a delight to eat!!

Preparation Time: 60 minutes
Servings: 12–24
Equipment: sheet pan, 2 small mixing bowls, mixing spoon, paper towels, airtight container

Ingredients:

* 15 oz. can chickpeas (garbanzo beans) or 2 cups cooked chickpeas
* 1 tsp. ground cinnamon
* 2 tsp. coconut oil, liquid
* 1 tbsp. honey

Directions:

1. Preheat the oven to 375°F.
2. Drain and rinse the chickpeas well, then pat them dry with paper towels.
3. In a small bowl, mix together the coconut oil and cinnamon.
4. Add the chickpeas into the bowl and stir until well coated.
5. Spread the chickpeas onto a baking sheet and bake them for 35–40 minutes, or until they're crunchy and no longer soft in the middle. (I shake the pan every 10 minutes.)
6. Place the hot chickpeas in a small bowl and coat them with honey, mixing it evenly.

7. Spread them back onto the baking sheet and allow them to dry and cool.
8. Store them in an airtight container at room temperature.

Trippy Fish

This recipe was originally for cats, but dogs loved it, too! Oddly enough, catnip makes cats act wild, yet it calms dogs...trippy right?

Preparation Time: 10 minutes
Servings: 6–10
Equipment: Ziploc bag

Ingredients:

* 12 oz. dried anchovies
* ¼ cup catnip
* Ziploc bag

Directions:

1. Empty the anchovies and catnip into the Ziploc bag.
2. Shake well until the catnip touches all of the anchovies.
3. Seal the bag and enjoy as necessary!

Dog-Friendly, Classic Peanut Butter Cookies

Your dog will love these super-easy peanut butter cookies. They're full of peanut butter flavor, with the perfect balance of salty and sweet, and a wonderfully crunchy, melt-in-your-mouth texture. This cookie is a classic for a reason!

Preparation Time: 30 minutes
Servings: 32
Equipment: mixing bowl, electric hand or stand mixer, fork, sheet pan

Ingredients:

* ½ cup *no sugar* peanut butter (literally just ground peanuts and oil)
* 1 tbsp. honey
* 1 egg
* 1 cup water
* 3 cups chickpea flour

Directions:

1. Preheat the oven to 375°F.
2. Cream the peanut butter, water, honey, and egg together in a bowl.
3. Stir in the flour until you achieve a dough texture. Your dough should not stick to your fingers. If it does, add ⅛ of a cup of flour.
4. Roll the dough into 1-inch balls and put them on a baking sheet.
5. Flatten each ball with a fork, making a crisscross pattern.
6. Bake them for about 10–12 minutes, or until the cookies begin to brown.
7. Allow them to cool, then enjoy!

Double Carob Cake

This is one of the richest-flavored cakes in this book. It truly embodies the beaty of a chocolate cake...without the chocolate! This recipe can also be made as cupcakes!

Preparation Time: 30 minutes
Servings: 6–12
Equipment: 2 mixing bowls, electric hand or stand mixer, round cake pan, mixing spoon

Ingredients:

* 1 cup all-purpose flour
* ¼ cup coconut oil
* ½ cup honey
* ½ cup water
* 1 tsp. baking powder
* ¼ cup carob powder
* 1 egg

For the Icing:

* 16 oz. plain Greek yogurt
* ¼ cup carob powder

Directions:

1. Preheat the oven to 350°F.
2. Combine the dry ingredients (except the carob powder) in a stainless-steel bowl, then add the coconut oil and water.
3. Using your mixer, start mixing at speed 1, then increase to speed 2 and mix for 1 minute.
4. Add your egg and ¼ cup of the carob powder to the bowl and up the mixer to speed 3 for about 30 seconds.
5. Turn it up to speed 4 and beat the mixture for about 1 minute, or until it's completely smooth and evenly mixed.

6. Pour the batter into a greased 8- or 9-inch round baking pan.
7. Bake it for 30–35 minutes, then test it with toothpick.
8. Remove it from oven and allow it to cool.

For the Icing:

1. In a small bowl, combine the Greek yogurt and ¼ cup of carob powder.
2. Stir it by hand until it's thoroughly mixed and reaches the icing texture.
3. Spread it on the cooled cake.

Plain Frozen Yogurt Recipe

Preparation Time: 3 hours
Servings: 4–8
Equipment: steel mixing bowl, mixing spoon

Ingredients:

* 32 oz. plain yogurt
* ½ cup honey

Directions:

1. In a large metal bowl, combine the yogurt and honey
2. Stir the mixture until it's thoroughly combined.
    * This is the most important step. You want the honey to mix well, as that is what keeps the frozen yogurt soft.
3. Cover the bowl and put it in the freezer.
4. Allow it to freeze overnight.
5. Scoop and serve!

Kale Pretzels

Pennsylvania Dutch pretzels are very common in Maryland where I live. They're probably one of the most consumed snacks in this region, and definitely one of my staples. This is a gluten-free recipe with the added superpowers of kale. The smell and flavor are irresistible for most dogs, and they make a great treat after a day of being a good pup!

Preparation Time: 40 minutes
Servings: 10–12
Equipment: 2 mixing bowls, sheet pan, brush, mixing spoon (wooden preferred), parchment paper or silicone baking mat

Ingredients:

* 1½ cups water (room temperature)
* 1 packet active dry or instant yeast (2¼ tsp.)
* 1 cup chopped kale
* 1 tbsp. honey
* 3¾–4 cups chickpea flour, plus more for work surface
* 2 tbsp. aluminum-free baking powder
* 1 egg

Directions:

1. Preheat the oven to 400°F.
2. Whisk the honey and water in a bowl.
3. Slowly add 3 cups of flour to the bowl, 1 cup at a time.
4. Add your baking powder to the bowl and mix with a wooden spoon (or dough hook attached to stand mixer) until the dough is thick. Add ¾ of a cup more of the flour until the dough is no longer sticky. If it's still sticky, add ¼–½ a cup more, as needed.
5. Turn the dough out onto a floured surface, then knead the dough for 3 minutes and shape it into a ball.

6. Return the dough to the bowl and add kale. Continue kneading the dough until the kale is evenly distributed.

7. Break off 2 oz. pieces of dough and roll them into smaller dough balls.

8. Line your baking sheet with parchment paper or silicone baking mats.

    * Silicone baking mats are highly recommended over parchment paper, because the edges of the paper may burn or can sometimes stick. If you're using parchment paper, lightly spray it with coconut oil spray before baking.

9. Roll each dough ball into a 12-to-14-inch rope. Take the ends and draw them together to form a circle.

10. Twist the ends, then bring them toward yourself and press them down into a pretzel shape. Repeat this step until you fill your pan.

11. In a small bowl, whisk the egg. Then brush the pretzels with the egg mixture.

12. Bake them for 12–15 minutes or until golden brown.

13. Allow them to cool before serving.

Jell-O Recipe

When I think of a fun snack, I always think of Jell-O! What most of us don't know is that the key ingredient in Jell-O is great for dogs with joint issues. Give 'em a little Jell-O every day to help lubricate the joints!

Preparation Time: 15 minutes
Servings: 2–4
Equipment: 8-by-8-inch glass dish, glass mixing bowl, saucepan

Ingredients:

* 4 cups coconut water
* 2 envelopes (¼ oz. each) unflavored gelatin
* 1 cup of your dog's favorite fruit, if desired (Heidi liked bananas and blueberries)

Directions:

1. Place 1 cup of coconut water in a glass bowl and sprinkle in the gelatin.
2. In a saucepan, bring the remaining 3 cups of coconut water to a boil.
3. Pour the boiling water over the gelatin-water mixture and stir until the gelatin dissolves completely.
4. Pour the mix into an 8-by-8-inch glass dish and add the fruit.
5. Allow it to cool in the refrigerator for 4 hours.
6. Cut it into squares and serve.

Sweet Potato Fries

These crispy, crunchy delights are great for dogs' hearts and digestive systems. Use them as training treats, like as a reward for not peeing on the floor. Whatever the reason, sweet potato fries are always in season!

Preparation Time: 4 hours
Servings: 10
Equipment: sheet pan, chef's knife, parchment paper

Ingredients:

* 2 large sweet potatoes

Directions:

1. Preheat the oven to 180°F.

2. Slice the sweet potatoes in half lengthwise, then again about ⅛- to ¼-inch thick. For smaller dogs, slice the sweet potatoes into rounds starting from one end and working to the other. For larger dogs, slice them lengthwise to make larger slices. Here's a look at my slices. They do shrink during the cooking process, so if you cut them thinner, they'll cook faster.

3. Line your baking sheet with parchment paper and lay your sweet potato pieces evenly onto it.

4. Bake for 3–4 hours, checking them frequently. If you want chewier fries, put them in for less time. For crispy, crunchy fries, bake them longer (but don't burn them).

5. Remove them from heat and allow them to cool before serving.

Wanna try something cool? Try substituting sweet potato for banana or green beans!

Yogurt Parfait Recipe

What's the best way to start your day? A healthy and delicious yogurt parfait! For those of you who feed your dog twice a day, replace the first meal with this amazing dish and restore their digestive system's good bacteria!

Preparation Time: 10 minutes
Servings: 1–2
Equipment: mixing bowl, small glass dish

Ingredients:

* Plain Greek yogurt
* Oatmeal
* Sliced banana
* Mixed berries (blueberries and strawberries)
* Flax seeds

Depending on the size of your dog, the ingredient measurements may vary. For a small dog, use about ¼ of a cup of yogurt for the entire recipe. For a larger dog, start with one cup.

Directions:

1. In a small dish, layer your ingredients, starting with a scoop of yogurt.
2. Add a layer of fruit, then add another layer of the yogurt.
    * (You can just do two layers or add a third.)
3. Add the final layer of fruit, then sprinkle some flax seeds on top.
4. Enjoy!

Halloween Peanut Butter Cups

I would say that out of all Halloween candies, the most valuable is the peanut butter cup. Kids would trade any and all candy for these, and I understand why. The combination of textures and flavors is unmatched. That's how your dog will feel about these peanut butter cups. If you're throwing a Howl-O-Ween Pawty, these are a must!

Preparation Time: 20 minutes
Servings: 12
Equipment: sheet pan, mini paper baking cups, mixing bowl, offset spatula or small spoon

Ingredients:

* 1½ cup unsweetened carob powder
* 1½ cup coconut oil
* 1 cup peanut butter (no sugar or xylitol, ground peanuts and oil only)

Directions:

1. In a small saucepan, melt the coconut oil over low heat until it liquifies, then remove it from the heat.
2. Add the carob powder into a small bowl. Slowly stir in the coconut oil until it's full mixed and looks like melted chocolate, then set the mixture aside.
3. Set 18 mini paper baking cups on a baking sheet.
4. Spoon 1 teaspoon of melted carob into each cup. Use an offset spatula or a small spoon to spread the carob slightly up the sides of each cup, and make sure you have an even base at the bottom of each.
5. Let them cool at room temperature for about 15–20 minutes until they're mostly solid.

6. Put your peanut butter in a sandwich bag.
7. Cut one corner of the plastic bag and pipe 1 teaspoon of peanut butter into the center of each cup.
8. With a very lightly moistened finger, tamp down on the peanut butter to make it flat and even, but leave a bit of space between the peanut butter and the edge of the paper cup (you should be able to see a ring of carob peeking out from below the peanut butter).
9. Spoon 1 teaspoon of carob onto the top of each cup. Use the spatula or a small spoon to spread the carob evenly over the top and down the sides of the cup.
10. Refrigerate them until they're solid (about 30 minutes).

Pumpkin Pie Recipe

Remember MoonPies? This recipe is my take on those MoonPies we all used to love. Filled with pumpkin meat, these are an awesome source of fiber and antioxidants!

Preparation Time: 35 minutes
Servings: 12
Equipment: sheet pan, brush, 2 mixing bowls, mixing spoon, 4- or 6-inch diameter round cookie cutter, teaspoon, fork, parchment paper or silicone mat

Ingredients:

* 1 can pumpkin meat, or 1 16 oz. fresh pumpkin
* ¼ cup honey
* 1 cup water
* 4 cups chickpea flour
* 2 eggs

Directions:

1. Preheat the oven to 350°F.
2. In a small bowl, whisk together the water, 1 egg, and honey.
3. Slowly stir in 3 cups of flour (but save that last cup). Stir until the mixture becomes a dough consistency. If it sticks to your fingers, add about ¼ of a cup from that extra cup of flour.
4. Form the dough into a ball.
5. Use the remaining flour to sprinkle your surface.
6. Flatten the dough ball onto your floured surface, and use rolling pin to roll the dough to ⅛-inch thickness.
7. Use a 4- or 6-inch diameter round cookie cutter and cut out 12 circles. If you're using the 4-inch cutter, you will yield more pies.

8. Line your baking sheet with parchment paper or use a silicone mat.
9. Lay the circles on the baking sheet, and using a teaspoon, put a level spoonful of pumpkin in the center of each circle.
10. Fold the circles in half, so they look like tacos laying on their sides.
11. With a fork, press the edges of the pies to seal them.

12. In a small bowl, whisk the other egg and brush the pies with the egg mixture.
13. Bake the pies for 15 minutes, or until golden brown.
14. Cool and serve.

Homemade Chicken or Beef Jerky Dog Treats

In case you didn't know, it's really easy to make dehydrated treats at home. Dogs love them, and they're great for several reasons. Both puppies and adult dogs love to chew, and these snacks take a longer time to eat than cookie-style treats. As usual, treats are treats, and should not be made as a substitute for meals.

Most people use a dehydrator to make these types of treats, but your oven can work just as well. In order to safely dehydrate meat at home, it has to be cooked through and reach an internal temperature of 165°F for poultry, and 145°F for beef or fish. Some dehydrators have a "meat" setting or a temperature dial to ensure that this is met. Since your conventional oven may not have this feature, follow the temperature guidelines in the recipes.

Preparation Time: 3–6 hours
Servings: 10–20
Equipment: dehydrator or regular oven, chef's knife, sheet pan, parchment paper

Ingredients: When making chicken jerky, I use chicken breast tender cutlets, but you can also use boneless, skinless chicken breasts. Be sure trim off any fat and slice the breasts with the grain into ⅛- to 3/8-inch-thick pieces. The same goes for beef liver or pizzles. Chicken cuts tend to dry out faster than liver or pig ears. This method works for all options, but keep a closer, more watchful eye on the chicken and pig ears.

Directions:

1. Preheat the oven to 185°F.

2. Lay out your proteins on your dehydrator trays. If you're using an oven, use parchment paper to line your baking sheet and place the protein cuts on the baking tray.

3. Dry the meat for 3–12 hours. The thicker your slices, the longer they will take. The middle should be dry and moisture free, and a light brownish color.

4. To make sure all of the bacteria has been cooked out and your homemade dog treats are safe to eat, turn the oven up to 275°F for the last 10 minutes.

Quick Tip! If you have a smaller, older dog with fewer teeth, you can use their canned dog food for this treat. Put the canned food in a piping bag and press them onto the baking sheet.

Pumpkin Frozen Yogurt Recipe

Preparation Time: 3 hours
Servings: 4–8
Equipment: steel mixing bowl, mixing spoon

Ingredients:

* 32 oz. plain yogurt
* 1 can pumpkin meat (real pumpkin, not pumpkin pie filling!!)
* ½ cup honey

Directions:

1. In a large metal bowl, combine the yogurt, honey, and pumpkin.
2. Stir the mixture until it's thoroughly combined.
    * This is the most important step. You want the honey to mix well, as that's what keeps the frozen yogurt soft.
3. Cover the bowl and put it in the freezer, allowing the mix to freeze overnight.
4. Scoop and serve!

Mixed Berry Frozen Yogurt Recipe

Preparation Time: 3 hours
Servings: 4–8
Equipment: steel mixing bowl, mixing spoon, food processor

Ingredients:

* 32 oz. plain yogurt
* 16 oz. frozen mixed berries
* ½ cup honey

Directions:

1. In a large metal bowl, combine the yogurt and honey. Stir well.
2. Put the frozen berries in the food processor until they're fully blended and a beautiful purple color.
3. Pour the blended berry mix into the yogurt and honey mix and stir them thoroughly.
4. Cover the bowl and put it in the freezer overnight.
5. Scoop and serve!

Peanut Butter and Banana Frozen Yogurt Recipe

Preparation Time: 3 hours
Servings: 4–8
Equipment: steel mixing bowl, mixing spoon, food processor

Ingredients

* 32 oz. plain yogurt
* 4 bananas
* ½ cup natural peanut butter
* ½ cup honey

Directions:

1. Peel the bananas and grind them in the food processor.
2. In a large metal bowl, combine the yogurt, honey, and the banana mix.
3. Stir mixture until thoroughly combined, then add the peanut butter and lightly stir again.
4. Cover the bowl and put it in the freezer overnight.
5. Scoop and serve!

Bacon and Molasses Frozen Yogurt Recipe

Preparation Time: 3 hours
Servings: 4–8
Equipment: skillet, steel mixing bowl, mixing spoon, chef's knife, paper towel

Ingredients:

* 32 oz. plain yogurt
* 4 oz. low-sodium bacon
* ½ cup molasses

Directions:

1. In a skillet, cook the bacon over medium heat until it's crispy but not burned.
2. Remove it from heat, but do not throw away the bacon grease!!!
3. Remove the bacon from the pan and dry it on a paper towel. Once the grease cools down, pour it into a mixing bowl.
4. Add the yogurt and molasses to the bowl, then mix well.
5. Chop the bacon into small pieces and fold them into the mix.
6. Cover the bowl and put it in the freezer overnight.
7. Scoop and serve!

Carrot Cake Recipe

Preparation Time: 45 minutes
Servings: 6-12
Equipment: 2 mixing bowls, electric hand or stand mixer, mixing spoon, 9- or 10-inch tube pan or a brownie pan, coconut oil

Ingredients:

* 1 cup molasses
* 2 cups chickpea flour
* 2 tsp. ground cinnamon
* 1 tbsp. baking powder
* 1 cup olive oil
* 3 eggs
* 1 lb. baby carrots

For the Icing:

* ½ cup peanut butter (no sugar or xylitol)
* 1 lb. Greek yogurt

Directions:

1. Preheat the oven to 350°F.
2. For a full cake, grease one 9- or 10-inch tube pan with coconut oil. For mini cakes, use a brownie pan.
3. Combine the molasses, flour, cinnamon, baking powder, and oil in a bowl.
4. With an electric mixer, beat in the eggs one at a time.
5. Stir in the carrots.
6. Pour the batter into the prepared baking pan or brownie pan.
7. Bake for 20–25 minutes, then remove it from the oven.

For the Icing:

1. Mix the peanut butter and yogurt in a bowl.
2. Spread the mix evenly over the top of the cake(s) and top it with halved baby carrots.
3. Enjoy!

Carob-Dipped Strawberries

The first time I made these was for a promotional event. A pair of the other vendors was a couple, of which the wife was allergic to chocolate. Her husband told me that chocolate was her favorite thing, but she found out she was allergic over thirty years ago. I assured her that carob was not going to cause a reaction, so she tried the strawberries. She broke into tears after one bite, as it had been years since she tasted anything like chocolate. I'm not allergic, and I have to admit...the flavor of these would make a dog shed a tear!

Preparation Time: 20 minutes
Servings: 10–15
Equipment: saucepan, cold plate, mixing spoon

Ingredients:

* ¾ cup coconut oil
* ¾ cup carob powder
* 10–15 fresh strawberries
* Xanthan gum powder

Directions:

1. Put your plate in the freezer and allow it to get cold.
2. Add the coconut oil to a saucepan and turn the heat to low. Stir it until it liquefies, then remove it from the heat.
3. Add the carob powder and a pinch of xanthan gum to the saucepan and stir.
4. Pour the mixture into a small bowl.
5. Remove the plate from the freezer and get your strawberries while you're there.
6. Dip the strawberries in the carob mixture and set them on the cold plate.

7. Return the plate to the freezer for 3 minutes, then move it to the refrigerator.
8. Eat and enjoy!

Mini Cheesecake

Cheesecake is something that no one ever gives a dog. It's usually got tons of sugar and cream cheese. This version, however, has none of that—all of the bad stuff was replaced with good stuff! These are great for a Fourth of July dog party!!

Preparation Time: 45 minutes
Servings: 12
Equipment: muffin pan, rolling pin, brush, whisk, 2 mixing bowls, mixing spoon, 4-inch diameter round cookie cutter, cooking oil

Ingredients:

* ¼ cup honey
* 1 cup water
* 4 cups chickpea flour
* 2 eggs
* 16 oz. Greek yogurt
* 8 oz. tapioca starch
* 8 oz. mixed berries
* 1 kiwi fruit

Directions:

1. Preheat the oven to 350°F.
2. In a small bowl, whisk together the water, 1 egg, and honey.
3. Slowly stir in 3 cups of flour (but save that last cup). Stir until the mixture become a dough consistency. If it sticks to your fingers, add about ¼ of a cup from that extra cup of flour.
4. Form the dough into a ball.
5. Use the remaining flour to sprinkle your surface.
6. Flatten the dough ball on your floured surface and use the rolling pin to roll the dough to ⅛-inch thickness.

7. Use a 4-inch diameter round cookie cutter and cut 12 circles out of the dough.
8. Spray your muffin pan with cooking oil.
9. Lay the circles in the muffin pan and press them in place like pie shells.
10. Bake the shells for 20 minutes, then remove them from the heat and allow them to cool.

For the Icing:

1. In a small bowl, mix the tapioca starch and yogurt.
2. Fill the shells with the yogurt mixture.
3. Top them with fruit and drizzle ½ a teaspoon of honey over them.
4. Serve and enjoy!

Mom's Salmon Pie with Portobello Mushrooms (for humans only!)

When Christmas came around, before the family feast, there were all those gift boxes of candy and baked goodies in the month leading up to the big day. Of course, we shared the love with the dog—it was the season! And besides, she loved being treated! After eating all that turkey for Thanksgiving, we usually had boring turkey again for Christmas. However, my mom had the idea to start using salmon for Christmas dinner instead. Baked salmon was a healthier choice than turkey, and Dad's doctor said he should start eating more fish for the healthy fat and protein. The first time she just roasted the salmon, but then she started making a neat dish: salmon pie with portobello mushrooms in a Dijon white wine sauce. Poor Heidi got none of that because we ate it all and there were some ingredients she couldn't have. My mom continued to make this meal as our Christmas staple every year. Heidi never got to try that dish, but here's the recipe to Mom's salmon pie that you can serve to your *human* family!

Preparation Time: 45 minutes
Servings: 5 to 7 humans
Equipment: baking dish, chef's knife, cutting board, skillet or cast iron pan

Ingredients:

* 2 tbsp. butter
* 1 leek, finely chopped
* 2 tbsp. flour
* ¾ cup cream
* ¼ cup dry white wine
* ½ tsp. chives, minced (or dried)
* 3 tbsp. Dijon mustard
* 1 tbsp fresh tarragon, chopped
* 1 tsp. salt (or to taste)
* 3–4 small salmon fillets, skin removed
* 1 lb. portobello mushroom
* Frozen puff pastry

Directions:

1. Preheat the oven to 350°F.
2. Wash the chopped leeks thoroughly and let them dry.
3. Over medium heat, melt the butter in a skillet or cast iron pan. Add the leeks and sauté them until softened.
4. Add the flour, then the cream, and allow the sauce to thicken.
5. Stir in the wine, chives, and mustard. Bring the heat down to low and let it simmer for about 5 minutes.
6. Add tarragon and continue to let it simmer. Season the mixture to taste with salt or more mustard.
7. Cut the salmon into chunks, making sure to remove any bones.
8. Cut the portobello mushroom into thin slices and set them aside.
9. If using a cast iron pan, drape the puff pastry over top and let the crust move up the sides of the pan.
10. Cover the dough with the mushroom pieces, then add the salmon chunks.
11. Transfer the contents of the pan to the baking dish and place the crust on top.
12. Bake according to puff pastry directions.
13. Allow it to cool for 10–15 minutes after removing from the oven.

Acknowledgments

I would like to dedicate this book to my mother, who taught me to love my creative side and believed in me even when I did not believe in myself. To my brother Terry, who showed me that, through faith in my vision, I could achieve anything I set my mind to. To my sister Monique, who's been my biggest cheerleader since I hit the planet. To my boys Cade and Max, who love me for my crazy ideas and give me purpose. To Meredith for being my rock through all of this and not thinking I am insane for going down this path and for loving me for who I am. Last and not least, to my father for instilling a sense of pride and unstoppable motivation in everything I do.

And of course, Greta, for using me as a vessel to make the world a better place for dogs.

I love ALL OF YOU!

About the Author

Chef Kevyn Matthews got his start in the culinary arts at the young age of ten by reading his mother's cookbooks and experimenting at home. After graduating high school, he found work in the restaurant industry as a cook, and then moved on to a career in catering. Eventually, Chef Kevyn moved to New York to pursue his dream of being a musician but found himself back in the kitchen. He has worked with some of the best chefs in the world, including José Andrés, and has cooked on set for Danny Aeillo, Jane Curtin, and Heidi Klum. While in New York, Kevyn found a puppy that he fell in love with and named her Greta. Remembering the food related health issues his childhood dogs dealt with combined with the love he had for Greta inspired Chef Kevyn to study and create the healthiest meals possible for his new pup. Soon, others were reaching out to Chef Kevyn for nutritious meals for their dogs, and his new career as The Dog Chef took off. Since becoming The Dog Chef, Kevyn has been featured on Animal Planet, Voice of America, DOGTV, and most recently on *Pet Tales* with Greta Van Susteren. The Dog Chef has also had great success with his dog restaurant in Baltimore, which showcases all of his awesome creations.

Printed in the USA
CPSIA information can be obtained
at www.ICGtesting.com
JSHW012010140824
68134JS00023B/2357